ALLZEIT WACH 1842

W. Künzel M. Kirschbaum (Eds.)

OXYGEN: Basis of the Regulation of Vital Functions in the Fetus

With 84 Figures and 16 Tables

Springer-Verlag
Berlin Heidelberg New York
London Paris Tokyo
Hong Kong Barcelona
Budapest

Professor Dr. med. Wolfgang Künzel
Gf. Direktor der Frauenklinik und Hebammenschule
der Justus-Liebig-Universität
Klinikstraße 32, 6300 Gießen
Bundesrepublik Deutschland

Dr. Dr. med. Michael Kirschbaum
Frauenklinik der Justus-Liebig-Universität
Klinikstraße 32, 6300 Gießen
Bundesrepublik Deutschland

ISBN-13: 978-3-642-77471-3 e-ISBN-13: 978-3-642-77469-0
DOI: 10.1007/978-3-642-77469-0

Library of Congress Cataloging-in-Publication Data
Oxygen: basis of the regulation of vital functions in the fetus / W. Künzel, M. Kirschbaum (eds.). p. cm.
Includes bibliographical references and index.

1. Fetal anoxia – Congresses. 2. Oxygen – Physiological transport – Congresses. 3. Fetus – Respiration and cry – Congresses.
4. Maternal-fetal exchange – Congresses. I. Künzel, W. (Wolfgang), 1936-. II. Kirschbaum, Michael, [DNLM: 1. Fetal Anoxia – congresses. 2. Fetus – physiology – congresses, 3. Maternal-Fetal Exchange – congresses. 4. Oxygen – physiology – congresses. 5. Oxygen Consumption – physiology – congresses. WQ 210.5 098]
RG629.A55099 1992
618.3'2 – dc20
DNLM/DLC
for Library of Congress 92-20354
CIP

Softcover reprint of the hardcover 1st edition 1992

23/53145/5 4 3 2 1 0 – Printed on acid-free paper

Preface

One of the main sources of energy in human life is the steady supply of oxygen. Permanently high or low concentrations of oxygen are, however, hazardous and incompatible with life. Little is known about where the critical threshold lies and which regulatory and adaptive mechanisms are implemented in hazardous situations to ensure the continuation of human life.

From the very beginning the fetus in utero is in an environment of low oxygen partial pressure (pO_2). Adaptive mechanisms work to guarantee its development and oxygen supply. These regulatory mechanisms are, however, not well understood. Does the fetus sense its own oxygen need and, if so, how is this regulated and how does it interact with nutritional factors. Is fetal growth adjusted to the reduced oxygen supply in dangerous situations, and is the initiation of labor a result of a fall in pO_2 during gestation? It may be speculated that many cases of obstetric pathology are related to fetal oxygen deprivation which may result in fetal death.

This volume is a collection of papers presented at a symposium which was held on 14 November 1989 to mark the 175th anniversary of the Frauenklinik in Giessen, FRG. The participants came from various fields of research; they included biochemists, biologists, physiologists, obstetricians, pediatricians, and anatomists, all of whom devote their investigations to one subject, namely oxygen as a basis of the regulation of vital functions in the fetus. I would like to thank all of them for coming to Giessen and contributing to this symposium.

I would also like to thank the many helpers for their assistance. Particular thanks are due to Dr. Michael Kirschbaum for organizing this meeting, Frau Jakobi and Frau Hedrich for their excellent secretarial work, Frau Achenbach for slide projection, and all the staff of the Frauenklinik who offered their help. Finally, I wish to thank Springer-Verlag for its courtesy and generosity in the process of seeing the manuscript through to press.

Giessen, June 1992 W. Künzel

Contents

I Oxygen Supply of the Fetus During Pregnancy and Birth

Abstracts

II Acute Oxygen Deficiency During Pregnancy and Birth. Influence on the Fetus

Abstracts

Abstracts

IV Chronic Oxygen Deficiency During Pregnancy: Impact of the Fetus

Abstracts

Contributors

Prof. Dr. J. G. Aarnoudse
Dept. of Obstetr. & Gynecol., University Hospital, Oostersingel 59
9713 EZ Groningen, The Netherlands

Prof. Dr. H. Acker
Max-Planck-Institut für Systemphysiologie, Rheinlanddamm 201
4600 Dortmund 1, Germany

Dr. K. Akagi
Lawson Research Institute St. Joseph's Health Centre,
268 Grosvenor Street
London, Ontario N6A 4V2, Canada

Priv.-Doz. Dr. B. Arabin
Klinikum Steglitz, Frauenklinik und Poliklinik,
Abt. f. Gynäkol. Endokrinologie, Hindenburgdamm 30
1000 Berlin 45, Germany

Prof. Dr. C. Bauer
Physiologisches Institut der Universität Zürich, Winterthurer Straße 190
8057 Zürich, Switzerland

Prof. Dr. R. Baumann
Institut für Physiologie, Universitätsstr. 31
8400 Regensburg, Germany

Dr. R. Baustädter
Doblerstraße 32
8399 Rotthalmünster, Germany

Dr. A. D. Bocking
Lawson Research Institute St. Joseph's Health Centre,
268 Grosvenor Street
London, Ontario N6A 4V2, Canada

Dr. R. H. Bödeker
Institut für Medizinische Informatik, Heinrich-Buff-Ring 44
6300 Gießen, Germany

Dr. N. M. Bolas
SMIS limited, Alan Turing Road, Surrey Research Park, Guildford
Surrey GU2 5YF, UK

Dr./Univ. Gent G. Braems
Univ.-Frauenklinik und Hebammenschule, Klinikstraße 32
6300 Gießen, Germany

Dr. A. N. Brooks
MRC Unit in Reproductive Biology, University of Edinburgh
Edinburgh, UK

Prof. Dr. A. M. Carter
Department of Physiology, University of Odense, J. B. Winsløwsvej 19
5000 Odense C, Denmark

Dr. E. Cetin
Univ.-Krankenh. Eppendorf, Frauenklinik, Martinistraße 52
2000 Hamburg 20, Germany

Professor Dr. J. R. G. Challis
Lawson Research Institute St. Joseph's Health Centre,
268 Grosvenor Street
London, Ontario N6A 4V2, Canada

Dr. James F. Clapp III
Case Western Reserve University, School of Medicine at MetroHealth
Medical Center, 2500 MetroHealth Drive
Cleveland, OH 44109-1998, USA

Professor Dr. G. S. Dawes
Nuffield Department of Obstetrics, John Radcliffe Hospital
Oxford OX3 9DU, UK

A. Detmer
Department of Physiology, University of Odense, J. B. Winsløwsvej 19
5000 Odense C, Denmark

Dr. K.-U. Eckardt
Institut für Physiologie I, Universitätsstraße 31
8400 Regensburg, Germany

Dr. R. FABER
Frauenklinik der Universität, Philipp-Rosenthal-Straße 55
O-7010 Leipzig, Germany

Dr. M. A. HANSON
Fetal and Neonatal Research Group, Dept. of Obstetrics & Gynecology, University College and Middlesex School of Medicine, University College London, 86-96 Chenies Mews
London WC1E 6HX, UK

Prof. Dr. M. A. HEYMANN
Cardiovascular Research Inst., Gynecol. & Reproductive Sciences, University of California, Box 0544, Room HSE 1403
San Francisco, CA 94143-0130, USA

Priv.-Doz. Dr. M. HOHMANN,
Univ.-Frauenklinik und Hebammenschule, Klinikstraße 32
6300 Gießen, Germany

Dr. P. HOPE
Special Care Baby Unit, John Radcliffe Hospital, Headington
Oxford OX3 9DU, UK

Prof. Dr. Dr. h.c. A. HUCH
Dept. für Frauenheilkunde, Klinik u. Polikl. f. Geburtshilfe, Frauenklinikstraße 10
8091 Zürich, Switzerland

Prof. Dr. Dr. h.c. R. HUCH
Dept. für Frauenheilkunde, Klinik u. Polikl. f. Geburtshilfe, Frauenklinikstraße 10
8091 Zürich, Switzerland

Dr. N. K. IVES
Special Care Baby Unit, John Radcliffe Hospital, Headington
Oxford OX3 9DU, UK

Dr. R. A. JACOBS
Lawson Research Institute St. Joseph's Health Centre, 268 Grosvenor Street
London, Ontario N6A 4V2, Canada

Dr. J. JELINEK
Milupa AG, Scientific Department, Bahnstraße 14–30
6382 Friedrichsdorf/Ts., Germany

Prof. Dr. A. JENSEN
Univ.-Frauenklinik und Hebammenschule, Klinikstraße 32
6300 Gießen, Germany

Dr. Dr. M. KIRSCHBAUM
Univ.-Frauenklinik und Hebammenschule, Klinikstraße 32
6300 Gießen, Germany

Dr. A. KRIETE
Institut für Anatomie und Zytobiologie, Aulweg 123
6300 Gießen, Germany

Prof. Dr. W. KÜNZEL
Univ.-Frauenklinik und Hebammenschule, Klinikstraße 32
6300 Gießen, Germany

Dr. P. KUMAR
Department of Physiology, University of Birmingham, Vincent Drive
Birmingham B15, UK

Prof. Dr. A. KURTZ
Institut für Physiologie I, Universitätsstraße 31
8400 Regensburg, Germany

Dr. U. LANG
Univ.-Frauenklinik und Hebammenschule, Klinikstraße 32
6300 Gießen, Germany

Prof. Dr. W. MOLL
Inst. für Physiologie, Universitätsstraße 31
8400 Regensburg, Germany

Dr. J. MOORCRAFT
Special Care Baby Unit, John Radcliffe Hospital, Headington
Oxford OX3 9DU, UK

Dr. K. H. NICOLAIDES
Dept. of Obstetrics & Gynecology, King's College Hospital,
Denmark Hill
London, SE5 9RS, UK

Dr. A. NIENARTOWICZ,
formerly: Inst. für Physiologie, Universitätsstraße 31
8400 Regensburg, Germany

Dr. J. G. NIJHUIS
University Hospital Nijmegen, Dept. Obstetr. & Gynecol.,
P.O. Box 9101
6500 HB Nijmegen, The Netherlands

Dr. H. PARISH
Bristol-Myers & Squibb, Donaustaufer Str. 378
8400 Regensburg, Germany

Priv.-Doz. Dr. U. POHL
Institut für Physiologie der Med. Universität, Ratzeburger Allee 160
2400 Lübeck 1, Germany

Dr. G. K. RADDA
MRC Biochemical and Clinical Magnetic Resonance Unit,
John Radcliffe Hospital, Headington
Oxford OX3 9DU, UK

Dr. B. RAJAGOPALAN
MRC Biochemical and Clinical Magnetic Resonance Unit,
John Radcliffe Hospital, Headington
Oxford OX3, 9DU, UK

Dr. B. S. RICHARDSON
Lawson Research Institute St. Joseph's Health Centre
268 Grosvenor Street
London, Ontario N6A 4V2, Canada

Dr. R. ROBEL
Frauenklinik der Universität, Philipp-Rosenthal-Straße 55
O-7010 Leipzig, Germany

Prof. Dr. K.-E. RUCKHÄBERLE
Frauenklinik der Universität, Philipp-Rosenthal-Straße 55
O-7010 Leipzig, Germany

Primarius Dr. R. RUDELSTORFER
Landesfrauenklinik Wels, Linzer Str. 89
4600 Wels, Austria

Prof. Dr. A. M. RUDOLPH
Cardiovascular Research Institute, University of California,
Box 0544, HSE 1403
San Francisco, CA 94143-0130, USA

Dr. A. Ruider
Peschelanger 9
8000 München 83, Germany

Dr. T. M. Smits
Dept. of Surgery, University Hospital, Oostersingel 59
9713 EZ Groningen, The Netherlands

Dr. A. Sue-Tang
Lawson Research Institute St. Joseph's Health Centre
268 Grosvenor Street
London, Ontario N6A 4V2, Canada

Prof. Dr. B. Viehweg
Frauenklinik der Universität, Philipp-Rosenthal-Straße 55
O-7010 Leipzig, Germany

Prof. Dr. G. Wegener
Institut für Zoologie, Universität Mainz, Saarstraße 21
6500 Mainz 1, Germany

Dr. B. A. Williams
Fetal and Neonatal Research Group, Dept. of Obstetrics & Gynecology, University College and Middlesex School of Medicine,
University College London, 86-96 Chenies Mews
London WC1E 6HX, UK

Dr. M. Winkler
Frauenklinik der Universität, Philipp-Rosenthal-Straße 55
O-7010 Leipzig, Germany

Prof. Dr. K. H. Wulf
Univ.-Frauenklinik und Hebammenschule, Josef-Schneider-Str. 4
8700 Würzburg, Germany

I Oxygen Supply of the Fetus During Pregnancy and Birth

Fetal Oxygen Supply – Historical Considerations

K.-H. WULF

"A history of fetal respiration: from Harvey's question (1651) to Zweifel's answer (1876)" is the headline of a report on research on fetal respiration given by Donald H. Barron on the occasion of a symposium in his honor at Bryn Mawr, Pennsylvania in 1976 (Fig. 1). Barron describes more than 200 years of speculations and dead-end roads from the discovery of the fetal blood circulation by Harvey to the proof of a diaplacental oxygen transfer by Zweifel.

The situation was as follows: In 1651 Harvey asked: "How does it happen that the foetus continues in its mother's womb after the seventh month? – it lives in health and vigour more than two months longer without the aid of respiration at all." In 1876, Zweifel answered: "The placenta is the lung of the fetus. The fetus possesses an oxygen metabolism of its own" (Fig. 2).

Fig. 1. Donald Henry Barron, Ph. D.

Fig. 2. Paul Zweifel (1848–1927)

In this chapter, the starting point will be the last statement, i.e., the answer. In 1876, in the *German Archive of Gynaecology,* Zweifel reported about the typical spectral absorption bands of oxyhemoglobin in the umbilical cord blood immediately after birth and before the first breathing movements of the child. This finding is the foundation of how we understand the gas exchange within the placenta. It is now a proven fact, and no longer speculation. Zweifel himself wrote:

"... with the complete isolation of the fetus from atmospheric air, the blood of the mother can be the only source of oxygen, and thus a direct proof of the placental transfer of oxygen is provided. It must pass from the one blood to the other through the epithelium of the chorionic villi and it is of interest to learn how fast such a movement can take place between qualitively different blood types."

The assumption of a diaplacental oxygen transfer and the evidence of oxygen in the fetal blood do not say anything definite about the oxygen requirement in the fetal tissues. Sufficient proof of an independent oxygen consumption of the fetus would be for example, arteriovenous differences in the oxygen content of the umbilical blood vessels. Accurate direct measurements of the true blood gas concentration were not possible in the previous century; one depended on indirect clinical observations such as heat production and blood color.

Important in this respect were evaluations of observed color differences between blood vessels. As early as 1799, Scheel had discovered light red blood (rubentum) in the umbilical vein, and dark red blood (purpureum) in the umbilical artery. These color divergences were not verified by later probes, and Zweifel was able to point out methodical mistakes. When handled properly, and avoiding circulation disturbances and heat loss, the blood of the umbilical vein was constantly of a lighter red than that of the umbilical artery. While experimenting, Zweifel also observed the phenomenon of the color first changing and then the original color being required during asphyxiation and reanimation of the pregnant ewe. Within a year, Zweifel's observation were confirmed by Pflüger's colleague Nathan Zuntz.

Zweifel found the different color shades between maternal and fetal blood striking, but he did not understand them. Scheel had already remarked that oxygenated blood in the umbilical vein was not redder than the venous blood of adults, and Zweifel maintained that the coloring of blood in the umbilical vein was not as light a red as in the maternal arterial vessels of the same caliber. Today we know that the color of blood is mainly dependent on the oxygen saturation deficit, and not on the content of oxygen. Lundsgaards und Van Slyke (1923) found a critical value for the O_2 deficit of about 6–7 vol%, or to make things simpler, an absolute content of reduced hemoglobin of 5 g%. If counts are higher, one has to anticipate cyanosis. This means for the placental gas exchange: deoxygenized blood from the umbilical artery will appear cyanotic; oxygenized blood from the umbilical vein can hardly be of a lighter red than the venous mixed blood in adults. With polyglobulia, even a small saturation deficit brings about a blue coloring of the blood, whereas with severe anemia even a high degree of deoxygenation will hardly reach the critical values.

For a long time the question as to which dimensions should be attributed to the oxygen consumption in the fetal organism remained open. Judging from the ability of the fetus to produce heat independently, Gusserow, Zweifel's teacher, assumed that the fetus has a considerable oxygen requirement, comparable to that of a newborn child. Zweifel provided evidence for this assumption. He found out that tolerance to hypoxia and the reanimation period of rabbit fetuses correspond to those of newborn animals. Consequently, the oxygen requirement should be similar.

We know today that the consumption of oxygen per minute in a term fetus is about 5 ml/kg weight. This corresponds to the gas exchange of an adult at rest or doing light physical work. The fetus works as well: it grows. The idea of oxygen consumption by the fetus for the sake of growth had already occurred to Zweifel. He wrote:

"... under the favorable conditions a mammal fetus enjoys as far as heat and food is concerned, all the energy of the oxygen metabolism goes into the development of the organs, so that in a relatively short time the astonishingly fast growth of the embryo is possible."

Analytic research on placenta gas exchange has continued throughout the twentieth century following the introduction of proper methods for blood gas

Table 1. Results of the first umbilical cord blood sampling using a Van Slyke manometer

Umbilical vein			Umbilical artery			Authors
SO_2	CO_2	CCO_2	SO_2	CO_2	CCO_2	
–	9.02	–	–	5.85	–	Blair-Bell et al. 1928
38	8.77	–	13	3.09	–	Haselhorst 1929
45	10.14	40.71	15	3.40	46.21	Haselhorst and Stromberger 1930
50	10.5	48.4	16	3.3	51.2	Eastman 1930, 1932
63	13.3	–	30	6.3	–	Eastman 1930, 1932

analysis. Progress in the field of medical science is completely dependent on new techniques and methods. The history of our awareness of how the fetus is provided with oxygen is a perfect example of this fact: Landmarks are set by biomedical innovations and techniques.

In 1917, Van Slyke drew attention to the Manometer, an instrument he had developed for the measurement of O_2 and CO_2 in the blood. In the late 1920s, for the first time umbilical cord blood sampling was performed with the Manometer, more or less simultaneous in Liverpool by Blair-Bell et al. (1928), in Hamburg by Haselhorst and Stromberger (1931), and in Baltimore by Eastman (1930) (Table 1). The results were similar, but surprising. Even within the arterialized blood of the umbilical vein, oxygen saturation was found to amount to about only 40%–50%, and in the umbilical artery to about 15%.

Our conceptions about gas exchange in the placenta were long determined by the very low oxygen levels of the umbilical cord blood. One spoke of a O_2 deficiency in utero which was said to cause an increase in anaerobic glycolysis in the fetus. It was in 1930 that Anselmino and Hofmann came out with a new idea, namely fetal adaption to high altitude. The distributive effect was immense and is best demonstrated by the slogan "Mount Everest in utero", a bon mot attributed to Sir Joseph Barcroft (Fig. 3). It is true that the O_2 pressure in utero corresponds to an altitude of 8000–9000 m. However, at this altitude the CO_2 pressure is low as well. The fetus, however, shows an arterial hypercapnia with values of about 45 mmHg.

The mechanism of the diaplacental gas exchange, that is, the question of what forces bring about the transfer, has not been explained yet. Two opinions exist and are each diameterical, opposed to the other:

1. The vitalistic conception of an active, secretory, energy-consuming biological effect of the trophoblast (Hofbauer 1905, Bohr 1909); and
2. the mechanistic explanation of a passive, physical process, comparable to diffusion (Schwartz 1858; Cohstein and Zuntz 1884; Huggett 1927).

Today we maintain that gas exchange, even in the placenta, takes place through diffusion. At least there is no theoretical or experimental necessity to

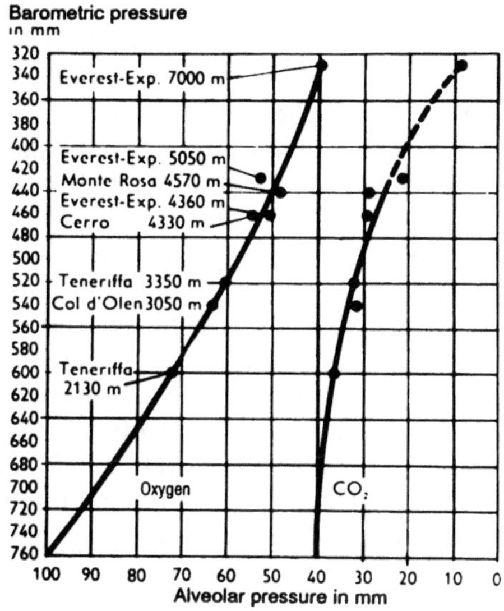

Fig. 3. Fetal adaption to high altitude

think of an active transfer of gas molecules. On the other hand there are a number of clear reports that indicate passive procedures. The mechanistic conception, based on voluntary transfer of molecules according to Fick's principle, is supported in the first place by evidence of adequate morphological structures characteristic of diffusion, and of sufficiently great pressure gradients for the blood gases O_2 and CO_2 through the placenta barrier. I will concentrate on the latter.

There were no adequate methods available to measure blood gas tensions directly until the middle of this century. We depended on experimental observations to trace the movement of gases, for instance, the so-called diffusion conversion during asphyxia, and on the calculation of pressure gradients with the help of oxygen dissociation curves. These operations were time-consuming and additionally led to false results.

The systematic research done with animals by A. S. Huggett (1927) was especially interesting. He initially measured the absolute gas concentration in maternal and fetal blood according to the Van Slyke method; then he graduated to construct in vitro oxygen dissociation curves with the help of the Barcroft saturator, and finally calculated the actual gas tensions. Huggett found a pressure divergency of 45 mmHg for O_2 and 18 mmHg for CO_2 between the arterial blood of the mother and the venous blood of the fetus. Using his method, which shows only relationships and pressure divergencies and is called *"Differentialtonometrie"*, he was always able to indicate sufficient diaplacental gradients for O_2 as well as CO_2, and vice versa. It is amusing to learn why Huggett preferred goats for his experiments. "The sheep used by Cohnstein is three to four times the price of a goat," he wrote.

Accurate methods to determine gas tensions directly did not exist before the middle of this century. In 1951, Heinz Bartels developed his dropping-mercury electrode, a potentiometric technique to determine the partial pressure of oxygen (pO_2) in just a few milliliters of blood. In the same year, Raczkowski, an assistant at the Department of Obstetrics at Kiel University, took the first direct oxygen pressure measurements within the umbilical cord (Raczkowski 1951). He presented his data at the 29th congress of the German Society of Gynecology in Munich in October 1952. His measured values were almost equal to the calculated values. Later evaluations, with improved instruments and using polarographic methods, came to similar results: A pO_2 of about 30 torr for the oxygenized mixed blood of the umbilical vein, and about 15 torr for the deoxygenized blood of the umbilical artery.

I have followed the concept of Raczkowski and continued in his line ever since 1955. We became interested in the pressure gradients for O_2 and CO_2 on both sides of the placental membrane. First we measured (Wulf 1964), simultaneously and directly, the oxygen pressure in the maternal and fetal vessels of the placenta using the new techniques. According to our research, the maternal blood enters the uterine artery with/at a mean pO_2 of 91 torr; in the uterine vein the pO_2 is still 33 torr. The fetal blood reaches the placenta by the umbilical artery with an pO_2 of only 10 torr, and leaves it after oxygenization in the umbilical vein with a pressure of 31 torr on average. On the arterial side of the placenta between the uterine artery and umbilical vein we found an uteroumbilical pressure difference for oxygen of about 60 torr, and on the venous side only 20 torr. In total the pressure divergency between incoming maternal and fetal blood measures about 80 torr.

It is remarkable how strongly oxygen pressure decreases in the maternal blood on its passage through the placenta, how little tension increases in the fetal blood, and how high the pressure gradients are. The situation within the lungs can serve as a comparison. During the gas exchange the oxygen pressure increases from 35 torr in the mixed venous blood of the lung arteries to about 90/95 torr in the arterialized blood of the veins. At the exchange membrane, we only observed an alveoloarterial O_2 pressure gradient of 5–8 torr on average. A fundamental difference between lung and placenta gas exchange exists insofar as in the lungs a gaseous and a liquid ingredient come into contact. In the placenta we have two fluid compartments that store oxygen with the aid of the hemoglobin. Part of the solution to the riddle is the s shape of the oxygen dissociation curves. The placental gas exchange for the maternal blood takes place in the upper, flat part of the curve, and for the fetal blood in its steep middle area. In short, the same amount of oxygen causes a great pressure reduction within the maternal blood, but only a small pressure increase in the fetal blood.

Sufficient pressure gradients alone do not completely explain why the gas exchange by diffusion is adequate. Marginal factors are important as well, expecially the magnitude of the perfusion rate. Oxygen has a high permeability coefficient. The placenta, as an organ of breathing, is limited in its performance more from convection than from diffusion. Decisive for the

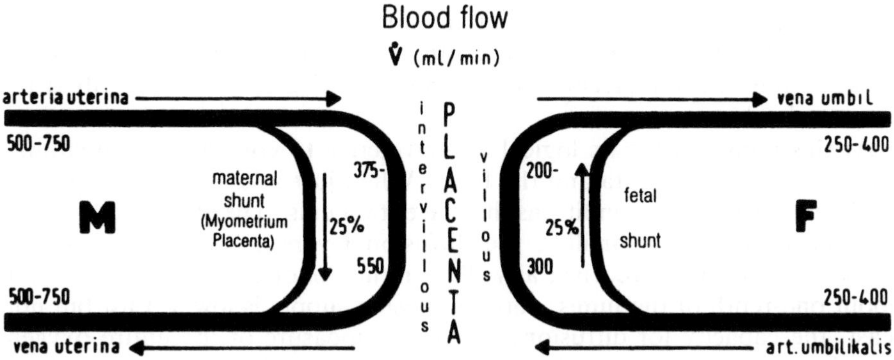

Fig. 4. The blood flow is measured on both sides of the placenta membranes to test the efficiency of gas exchange (Wulf 1964)

efficiency of the gas exchange is the flow on both sides of the placenta membranes. The total amount of blood rushing through the uterus comes to about 500–750 ml/min towards the end of pregnancy. About 75% of this quantity moves through the intervillous system, the remaining 25% are to be regarded as a shunt as far as gas exchange is concerned (Fig. 4). If the weight units of the uterus, placenta and fetus are taken into account, there is a stable average blood flow of 125–150 ml/min per kilogram. Continued controls during pregnancy showed the expected exponential increase of blood circulation due to growth, which corresponds with the increase of oxygen demand and of weight (Künzel and Moll 1972).

The blood circulation in the umbilical cord increases with progressing pregnancy as well. Immediately after birth one measures 250–400 ml/min. The average blood circulation per weight unit generally remains stable at 75–100 ml/min kg. On the umbilical side there is a shunt portion of about 25% as well (Dawes 1967).

The uteroplacental blood flow has a remarkable reserve as far as the oxygen demand of the fetus is concerned. It is true that a decrease in the uterine blood circulation always impairs the oxygen exchange, but that does not necessarily lead to an immediate reduction of the amount of O_2 exchanged. There are a number of compensatory mechanisms such as the extension of the arteriovenous difference in oxygen, the tachycardia of the fetus, and the reduction of the uneven distribution of the perfusion (homogeneous placenta). In addition, the efficiency of gas exchange is not only dependent on the given amount of blood circulation, but also on the arrangement of the maternal and fetal vessels and the direction of the blood flow. There is no rigid order in the coordination of the maternal and fetal blood vessels in the hemochorial placenta of the human. This differs from the countercurrent system in the labyrinth placenta of the rabbit. Obviously maternal blood does not run a predetermined course but finds its way through various lanes according to the actual pressure gradients (open flow system).

Bartels and Moll (1964) speak of a multivillous or multiserial system. Theoretically the efficiency of a multivillous placenta lies in the middle between the countercurrent and the concurrent data. This is verified by measured values.

At this point it appears logical and attractive to compare the breathing capability of the placenta and the lungs. When just visualizing the environment and the working conditions the placenta seems to be at a disadvantage. The diffusion area is smaller, the diffusion layers are longer, the blood pressure values in the receiver as well as in the donor system amount to just about one-tenth of the lungs, and the oxygen supply is lower. Last but not least, the capacity for diffusion, taken as a measure of the entire oxygen transfer, is ten times larger in the lungs. However, if one adjusts all these exchange rates to weight, or even better to the consumption of oxygen per minute in the tissues supplied, they are not so different (placenta and fetus $\dot{V}_{02}$ 15 ml/min ≈ 5 ml/kg per minute and lung V_{02} 300 ml/min ≈ 4 · 5 ml/kg per minute). The functioning of gas exchange in the placenta is absolutely adequate, effective, and sufficient. It is as delicate as lung breathing, and provided with only short-term reserves as well.

Today, more than 100 years after Paul Zweifel's epoch-making discoveries, his findings have stood the test of time, and the placenta is fully acknowledged as a breathing organ.

References

Anselmino KJ, Hoffmann F (1930) Die Ursachen des Ikterus neonatorum. Arch Gynäkol 134, 310

Barron DH (1978) A history of fetal respiration: from Harvey's question (1651) to Zweifel's answer (1876). In: Longo LD, Daniel DR (eds) Fetal and newborn cardiovascular physiology 1. Garland STPM Press, New York

Bartels H (1951) Potetiometrische Bestimmung des Sauerstoffdruckes im Vollblut mit der Quecksilbertropfelektrode. Theorie und Versuche. Pflügers Arch 254: 107

Bartels H, Moll W (1964) Passage of inert substances and oxygen in the human placenta. Pflügers Arch Gesamte Physiol Menschen Tiere 280D: 165

Blair-Bell W, Cunningham L, Jowett M, Millet H, Brooks J (1928) The metabolism and acidity of the foetal tissues and fluids. Br Med J I: 126

Bohr C (1909) Über die spezifische Tätigkeit der Lungen bei der respiratorischen Gasaufnahme und ihr Verhalten zu der durch die Alveolarwand stattfindenden Gasdiffusionen. Skand Arch Physiol (Berlin und Leipzig) 22: 221

Cohnstein J, Zuntz N (1884) Untersuchungen über das Blut, den Kreislauf und die Atmung beim Säugetierfoetus. Pflügers Arch XXXIV: 173

Dawes GS (1967) Physiology of the foetus in normal labour. In: Wood C, Walters AWW (eds) Fifth world congress of gynaecology and obstetrics. Butterworths, Australia

Eastman NJ (1930) Foetal blood studies. I. Oxygen relationship of umbilical cord blood at birth. Bull Johns Hopk Hosp 47: 221

Gusserow A (1872) Zur Lehre vom Stoffwechsel des Foetus. Arch Gynaekol 3: 24–270

Haselhorst G, Stromberger K (1930) Über den Gasgehalt des Nabelschnurblutes vor und nach der Geburt des Kindes und über den Gasaustausch in der Placenta. Z Geburtshilfe Gynäkol 98: 49

Haselhorst G, Stromberger K (1931) Über den Gasgehalt des Nabelschnurblutes vor und nach der Geburt des Kindes und über den Gasaustausch in der Placenta. Z Geburtshilfe Gynäkol 100: 48
Hofbauer I (1905) Grundzüge einer Biologie der menschlichen Placenta. Braumüller, Wien
Huggett ASG (1927) Foetal blood-gas tensions and gas transfusion through the placenta of the goat. J Physiol 62: 373
Künzel W, Moll W (1972) Uterine O_2 consumption and blood flow of the pregnant uterus. Z Geburtshilfe Perinatol 178: 108
Raczkowski HA (1953) Direkte Bestimmung des Sauerstoffdruckes im Nabelschnurblut des Neugeborenen. Arch Gynäkol 183
Wulf H (1964) The oxygen and carbon dioxide tension gradients in the human placenta at term. Am J Obstet Gynecol 88: 38–44
Zweifel P (1876) Die Respiration des Fötus. Arch Gynäkol 9: 291

*Oxygen and Animal Metabolism: General and Comparative Aspects**

G. WEGENER

Introduction

Oxygen is the most abundant chemical element in the earth's crust, it accounts for about 47% of the weight of the solid crust, and 88.8% of the hydrosphere. Bound in these inorganic compounds, oxygen cannot be used directly for animal metabolism. It is the free molecular dioxygen O_2 (less than 0.01% of the total oxygen in the earth crust) that is of eminent biological importance. Another form of molecular oxygen, the trioxygen ozone – concentrated in the ozone layer, 15–30 km above the earth's surface – is of indirect benefit because it protects organisms from radiation damage by absorbing most of the solar ultraviolet radiation.

Molecular oxygen is a major compound of the earth's atmosphere, but this has not always been so. Virtually all the free O_2 has been generated in the process of photosynthesis of green plants in which water is split and O_2 released as a waste product (see section on "Why aerobic metabolism?"). The present atmosphere (dry air at sea level) contains 20.95% O_2 by volume (23.13% by weight), whereas only about 1% O_2 is dissolved in pure water at 0°C and less at higher temperatures and in aqueous media (cf. Wegener 1988). The low capacity of water to dissolve O_2 has important biological consequences because most metabolic reactions occur in an aqueous environment. Only some groups of invertebrates, most notably the insects, transport O_2 in gaseous form thus making use of the high content of O_2 in the atmosphere and the much faster diffusion of O_2 in air than in water. Conveying oxygen in gaseous form to the organs is restricted to rather small animals because a delicate system of air-filled tubes (tracheae) is required, and this can only function over short distances. In insects, where this system has been evolved to great perfection, it contributes to sustaining the highest rates of aerobic metabolism encountered in the animal kingdom (for review and references, see Kammer and Heinrich 1978, Wegener 1990). In most groups of animals, however, O_2 is transported to the tissues by means of a circulatory system via respiratory pigments that bind oxygen with great capacity and affinity (for review, see Urich 1990). In vertebrates, hemoglobin functions as the oxygen

* Work from the author's laboratory has been supported by grants form the Deutsche Forschungsgemeinschaft, D-5300 Bonn

carrier, greatly increasing the O_2 carrying capacity of the blood (to about 20%, i.e. 20 ml O_2 in 100 ml blood in man).

Oxygen is involved in so many aspects of animal life that no attempt has been made to treat the topic comprehensively. Rather the discussion is restricted to some aspects of oxygen and energy metabolism including a brief account of adaptations for maternal-fetal gas exchange in ectothermic vertebrates. For a more comprehensive view, the reader is referred to the references, particularly the volume edited by Gilbert (1981).

A Historical Glimpse

Soon after the discovery of oxygen by C.W. Scheele (sometime between 1770 and 1773) and J. Priestley (1774), A. L. de Lavoisier set out to develop his oxidation theory, overturning the mystic phlogiston theory according to which combustible bodies contain phlogiston which is released in combustion. Working from his own experiments and taking careful note of the results of other investigators, Lavoisier correctly interpreted combustion and respiration as processes in which free oxygen is used up because it is combined with combustible matter: "Respiration is but a slow combustion of carbon and hydrogen, that is exactly like the combustion functioning in a lit candle" (for reference, see Gilbert 1981).

Before the existence of oxygen was known, the importance of air for life processes had already been clearly perceived. Robert Boyle, for one, was convinced that air or a part of it is required for combustion and life. He had observed that neither a flame nor an animal could survive in a vessel when this had been evacuated by an air pump, and in 1667 he expressed the idea: "There may be in the Heart and Blood a Vital Kind of Fire, which needs Air, as well as those Fires that are sensibly hot: Which may lessen the wonder, that Animals should not be able to live when robbed of Air" (Boyle 1668). Ancient ideas of Greek philosophy (e.g. Anaximenes, 585–526 BC) and medicine (e.g. Erasistratos, 300–240 BC) sound amazingly modern, according to which air contained a vital substance, the "pneuma" that was taken up by the lungs in the process of breathing, distributed via the blood to the organs where it was supposed to bring about life processes. The Greek word "anemos" (breath, soul) which is similar in meaning to "pneuma" is the source of the word animal (for review, Gilbert 1981).

Two observations relating oxygen and cell metabolism are connected with the name of Louis Pasteur and proved to have great impact on the development of modern biochemistry. Working with yeast, Pasteur (1861) observed that more yeast was produced per part of sugar if O_2 was present. The second observation (not quite correctly ascribed to Pasteur; for review, see Lagunas 1981) was that glucose is used up more slowly in the presence of oxygen that in its absence. This effect, named the "Pasteur reaction" by Warburg and later known as the "Pasteur effect", indicates the existence of an oxygen-dependent control mechanism governing the rate at which glucose is metabolized in living cells. Pasteur regarded yeast as the causal agent of fermentation and fermentation itself as a feature of life which did not seem approachable in terms of chemistry and physics (cf. Krebs 1972). Justus von Liebig, favoring chemical analysis, violently contested this view by stating that "causes cannot be seen with a microscope".

The Pasteur effect puzzled and challenged biochemists for decades. The missing links were provided by modern bioenergetics which couples oxygen consumption with the generation of biological energy in the form of adenosine triphosphate (ATP), i.e., the form of chemical energy required for executing physiological functions (see also section on "Why aerobic metabolism?"). In 1929 Lohmann detected ATP, and in 1941 Lipmann emphasized the universal role of "energy-rich phosphate bonds" for cell functions. But it was a whole century after Pasteur's observations, that the fundamental mechanism by which oxygen consumption and cellular functions are linked was revealed. Peter Mitchell's (1961) revolutionary proposal of a chemiosmotic mechanism of oxidative phosphorylation (i.e., ATP synthesis in the mitochondria) met with fierce resistance, but is now widely accepted. Electrons substracted from the food stuffs (oxidation) flow down the respiratory chain to O_2, and the energy available from this process is used to pump protons across the inner mitochondrial membrane thus forming a pH-gradient and a membrane potential (proton motive force). ATP synthesis is "driven" by the re-entry of protons through ATP-synthesizing enzymes (F_0F_1ATPases). These processes share a fundamental similarity in bacteria and in the ATP producing organelles (chloroplasts and mitochondria) of eukaryotes suggesting this form of energy transformation to be an early achievement in the evolution of life. How oxidative phosphorylation is regulated, i.e. how cellular work and respiration are coordinated is still not fully understood and an active area of research (for reviews Balaban 1990, Wegener et al. 1991).

At about the same time when Mitchell proposed the chemiosmotic mechanism of ATP production, phosphofructokinase (PFK) was definitely identified as the key element in the Pasteur effect. PFK had earlier on been suggested to be involved in the Pasteur effect (for review, see Krebs 1972) but its pivotal role as a pacemaker of glycolysis was comprehensible only after Passonneau and Lowry (1962) had shown that PFK activity was controlled by a set of key intermediates of energy metabolism: ATP at near-physiological concentrations inhibits PFK whereas metabolites that are formed upon ATP hydrolysis such as inorganic phosphate (Pi), adenosine diphosphate (ADP) and particularly adenosine monophosphate (AMP) counteract this inhibition. If oxygen is available many tissues utilize substrates other than carbohydrate. This leads to a further decrease in glycolytic flux (i.e., sparing of carbohydrate) which is important under conditions when carbohydrate is in short supply, for instance in starvation or during long term exercise. Oxidation of fatty acids or ketone bodies is supposed to bring about an increase in the content of citrate and this could, in part, be responsible for a decrease in PFK activity through an inhibitory effect of citrate on PFK (for review, see Newsholme and Leech 1983). So, about 100 years after Pasteur's work it was accepted that the Pasteur effect is a consequence of aerobic energy metabolism, based on allosteric feed back mechanisms, and mediated through ATP and other metabolites that reflect the cellular energy status and the availability of substrates (for review, see Krebs 1972; Racker 1980).

At this stage, regulation of glycolysis in animal tissues was considered by many authors to be a closed chapter. It came as a distinct surprise, when in 1980 fructose 2,6-bisphosphate, a substance previously unknown to exist in biological material, was detected as the most potent activator of phosphofructokinase in liver (Van Schaftingen et al. 1980). Fructose 2,6-bisphosphate was subsequently demonstrated in all mammalian cell types and is probably present in tissues from all animals. Fructose 2,6-bisphosphate is a potent activator of animal PFKs but, apart from liver, little is known about its physiological functions in the various tissues (for review, see Van Schaftingen 1987). Recent reports suggest that fructose 2,6-bisphosphate is also a signal for the integration and coordination of metabolic pathways in extrahepatic tissues. Fructose 2,6-bisphosphate and glycolytic flux are correlated in frog skeletal muscle during exercise (Wegener et al. 1990). In the flight muscle of the migratory locust fructose 2,6-bisphosphate is markedly decreased during prolonged flight when the locusts change from carbohydrate to fat as the major fuel (Wegener et al. 1986a, 1987). It is not known whether fructose 2,6-bisphosphate serves a similar function also in mammalian muscle during endurance exercise. But in starvation, another condition in which fat is the predominant fuel in muscle, fructose 2,6-bisphosphate is also decreased (French et al. 1988).

Some animals, most notably anoxia tolerant invertebrates, have recently been shown to decrease rather than increase the rate of glucose catabolism when O_2 is in short supply. The biological significance of this response is obvious: decrease of energy expenditure while aerobic ATP synthesis is precluded (see below). The mechanisms by which this effect, sometimes called the "negative Pasteur effect", is brought about are not, however, fully understood (for review, see Storey 1985; Wegener 1988).

Why Aerobic Metabolism?

The earliest forms of life evolved in an environment very different from the present biosphere. The primitive atmosphere was mildly reducing and did not contain free oxygen. Eukaryotic cells appeared only after early photosynthetic prokaryotes had turned the reducing atmosphere into an oxidizing one by releasing molecular oxygen (see Harold 1986; Rutherford 1989). At that time, life had already a long evolutionary history based on energy metabolism with oxidations performed anaerobically as dehydrogenations (by removing hydrogen together with its electron). Organisms have never abandoned this means of oxidation in their catabolic pathways. Aerobes added oxygen at the end of the scheme for "combustion" of the hydrogen (cf. Wald 1966). For a brief account of the evolution of energy metabolism, see Darnell et al. (1986), Alberts et al. (1989), Nelson and Taiz (1989). As mentioned above animal life evolved in a biosphere containing O_2, and all animals are primarily aerobes producing most of the ATP by membrane-bound ATP synthases in mitochondria containing respiratory chains with O_2 as final electron acceptor. This mode of energy production is not without hazards because O_2 can give rise to very toxic products (see section on "Hazards from oxygen"). So, what are the advantages (apart from recycling the hazardous O_2 to H_2O) that may explain the almost universal occurrence of aerobic energy production? Molecular oxygen combines two properties that are crucial for its function in energy metabolism. Firstly, the oxidative reactions have a high thermodynamic driving force, i.e., have a high potential for releasing energy or performing work. Secondly, O_2 reacts slowly unless activated by catalysts (enzymes). Indeed, oxyhydrogen gas (*Knallgas*) for instance, is quite stable at physiological temperatures, but if it reacts a large amount of energy is suddenly released with an impressive bang.

$$H_2 + \tfrac{1}{2}\,O_2 \rightarrow H_2O \quad \Delta\, G^\circ = -237.4\ \text{kJ}\ (= -56.7\ \text{kcal})/\text{mol}$$

It is even more impressive to realize that this reaction takes place, in a stepwise and fully controlled manner, in the mitochondria of living cells when the hydrogen derived from substrate molecules is oxidized to H_2O.

Compared with anaerobic metabolism aerobic energy metabolism produces a large amount of ATP, for instance

$$1\ \text{glucose} + 6\ O_2 + n\ \text{Pi} + n\ \text{ADP} \rightarrow 6\ H_2O + 6\ CO_2 + n\ \text{ATP}$$

(depending on how cytosolic reduced nicotinamide adenine dinucleotide, NADH, is shuttled into the mitochondria, the yield of ATP is $n = 36$ or 38, and of these, 34 or 36 ATP originate in the mitochondria, and 32 or 34 by oxidative phosphorylation).

In vertebrate tissue under anaerobic conditions, 1 mol glucose yields only two mols of ATP according to

$$1\ \text{glucose} + 2\ \text{ADP} + 2\ \text{Pi} \rightarrow 2\ \text{ATP} + 2\ \text{lactate} + 2\ H^+$$

In addition to a meager ATP yield, anaerobic metabolism has the disadvantage of being restricted to one substrate, carbohydrate, whereas aerobic ATP production can utilize all substrates. Moreover, aerobic metabolism produces harmless and readily excretable end products, H_2O and CO_2, from the main foodstuffs, carbohydrate and fat. In contrast, anaerobic glycolysis gives rise to lactate and protons which can upset cell metabolism and interfere with cellular functions.

Consequences of Anaerobic Metabolism

The fact that all animals are primarily aerobes does not imply that anaerobic energy production does not occur in animal tissues. Some specialized parts of an animal may well work anaerobically while the animal as a whole is aerobic.

In mammals, certain cells and tissues such as red blood cells, lens, parts of the retina, kidney medulla (and some tumors) are known to produce all or most of the ATP anaerobically, even in the presence of oxygen. Other organs take up and metabolize the anaerobic products of these cells.

Permanent anaerobic energy production of whole animals is rare. It is restricted to species adapted to environments very poor in oxygen, such as endoparasites in the intestinal tract of vertebrates (for review and references, see Wegener 1988). In the majority of animals anaerobic ATP production is not sufficient to maintain normal body functions. Therefore, if an animal encounters oxygen lack, the hypoxic or anoxic interval can be tolerated only for short periods and must be followed by an aerobic recovery period.

The actual reaction to anoxia, however, and the capacity for recovery from anoxia are remarkably different in different animals. Mammals, humans included, are extremely sensitive to anoxia: the function of their central nervous system (CNS) is lost within seconds, and 5 to 8 min of anoxia will lead to irreversible damage of the brain (for review and references, see Siesjö 1978; Wegener et al. 1986b; Wegener 1988). Some ectothermic vertebrates such as crucian carp, common frog, and turtles can cope with anoxia for extended periods, particularly at low environmental temperatures. Anoxia provokes coordinated reactions of the whole body and its organs that are aimed at sparing energy and maintaining the function of the CNS and circulation. These reactions include bradycardia, peripheral vasoconstriction, and redistribution of the blood flow to serve the vital organs such as brain and heart. A marked reduction in the overall metabolic rate is also observed under hypoxia. These mechanisms seem to have arisen early in evolution as they have been observed in all vertebrates studied so far, from fish to diving mammals (for review, see Wegener 1981, 1988), and also in embryos and fetuses (Jensen et al. 1987). Little is known about the mechanisms that control organ functions and metabolic rate under hypoxic conditions (for review, see Hochachka and Guppy 1987; Wegener 1988).

Oxygen and Embryonic Energy Metabolism in Ectothermic Vertebrates

The most common form of animal reproduction is oviparity, in which eggs are expelled from the maternal body with the embryo developing and hatching outside the body. In many species, eggs contain large amounts of yolk for the growing embryo. Yolk is rich in protein, yielding building blocks for body mass, and fat to provide energy. Fat is the preferential energy store because of its favorable energy to mass ratio. In contrast to carbohydrate, fat can be stored almost free of water, so that 1 g fat is equivalent to about 8 g glycogen (including tissue water). Producing eggs is a large but necessary investment because the embryo is cut off from external supply. It goes without saying that the embryo must use the stored material in the most efficient way, i.e., aerobically. Anaerobic energy production would be grossly inadequate. Even the offspring of intestinal parasites embark on aerobic metabolism in the early stages of their life cycles.

Viviparity has been evolved in many phyla of the animal kingdom, often independently in closely related forms. In vertebrates it is found in all classes, with the notable exception of the birds. Nutritionally, viviparity ranges from the mere retention of yolky eggs that are independent of the maternal organism (lecitrophy) to excessive matrotrophy, where the mother provides almost all the nutrients for the growing embryo. In terms of evolution the latter type is the more advanced. Irrespective of the trophic relationships, embryonic metabolism in viviparous species appears to be aerobic. In lecitotrophic species, the limited amount of stores requires a strictly aerobic metabolism of the embryo. Hence the more advanced matrotrophic forms probably originated from ancestors that already had some adaptations to maternal-fetal gas exchange.

Because viviparity has evolved so many times and in so many ecologically different animals, a wide diversity of adaptations is likely to exist. The few species that have been studied in some detail show organic, cellular, and molecular adaptations for maternal-fetal gas exchange. Much more is known about morphological and anatomical than about physiological adaptations. We will restrict the following discussion to ectothermic vertebrates and focus on a few examples. Viviparity is found in sharks, skates, and rays (Chondrichthyes, cartilagenous fish), in more than a dozen families of teleost fish, in urodeles and anurans and in many reptiles. A more comprehensive account is given in recent reviews (Amoroso 1968; Hogarth 1976; Wourms 1981; Seymour 1984; Yaron 1985; Korsgaard and Weber 1989).

Morphological and Anatomical Adaptations to Maternal-Fetal Gas Exchange

A placenta (or analogous structures) can be found in many viviparous vertebrates. In some sharks the distal portions of the yolk sac are in close contact or interdigitate with areas of the uterus. Both embryonic and maternal tissue is modified (vascularized) for facilitation of gas exchange. In skates and rays threadlike trophonemata are formed by the uterine wall, which may enter the esophagus of the fetus through the spiracles (for review and references, see Hogarth 1976; Korsgaard and Weber 1989).

Viviparous teleost fish do not gestate their embryos in modified parts of the oviduct (as other viviparous vertebrates do) but keep them within the ovary. The eggs develop either in the ovarian follicles (follicular gestation, as in the guppy, *Poecilia reticulata*, or the swordtail and platy, *Xiphophorus helleri*, and *X. maculatus*) or in the lumen of the hollow ovary (*ovarian gestation* as in the blenny, *Zoarces viviparus*). In some fish ovarian gestation follows follicular gestation, and some species show eccentric forms of gestation such as the seahorse (*Hippocampus*), where it is the male which receives the eggs from the female and becomes “pregnant”.

The anatomical adaptations to viviparity among fishes are manifold. Specializations of the epidermis, fin epithelium (e.g., in the seaperch, *Embiotoca*), gut, and yolk sac on the part of the embryo correspond to specializations of the epithelium of the maternal ovary: in many cases a follicular placenta is formed (for review, see Wourms 1981; Korsgaard and Weber 1989). In the guppy embryo, membranes surrounding the heart expand into a huge pericardial sac that extends over the head of the young thus forming a placenta-like structure for gas exchange. In the goodeid fish, *Ameca splendens* (Goodeinae), extraembryonic membranes form ribbons (trophotaenia), which are in direct apposition to the internal epithelium of the maternal ovary (Lombardi and Wourms 1985a, b). In some perches (Embiotocidae), specialized ovarian tissue intrude into the fetal gill cavities, thus facilitating the exchange of gases and nutrients (for review, see Wourms 1981).

Viviparity has been evolved in all extant orders of amphibia. It is most advanced in the limbless Apoda (caecilians). The young of *Typhlonectes,* a caecilian living in caves, have large external gills which serve as placenta. Little is known about the maternal-embryonic gas exchange in live-bearing salamanders such as *Salamandra atra,* which lives in the Alps and has a gestation period of about 1 year (2–3 years at great altitude), after which it gives birth to normally two young. True viviparity is rare in anurans and seems to occur only in the small toad, *Nectrophrynoides occidentalis,* which lives at altitudes of about 1200 m in tropical West Africa where wet and dry seasons alternate. No placental structures are formed, but gas exchange is certainly facilitated by the high vascularization of both the embryonic tail and the wall of the uterus (for review, see Wake 1980).

In reptiles, live-bearing is not found in turtles and crocodilians, but about one-fifth of the Squamata (lizards and snakes) are viviparous and have developed a wide range of adaptations to facilitate maternal-fetal exchange (for review, see Yaron 1985). Viviparous reptiles differ from all viviparous ectothermic vertebrates in that their placentae contain chorionic and allantoic elements homologous to those of mammals. The chorioallantoic placenta is the typical form of the reptilian placenta.

Physiological Adaptations to Maternal-Fetal Gas Exchange

Oxygen affinity of respiratory pigments can be adapted to environmental conditions and various mechanisms bring about these effects. Embryonic blood often shows a higher affinity for O_2 than adult blood. This holds true for many oviparous species, and viviparous vertebrates have almost invariably used this means to facilitate the transfer of O_2 from the maternal to the fetal circulation. Various mechanisms, singly or combined, can be used to this end. The most profound adaptation is the expression of unique fetal hemoglobins that have a higher intrinsic O_2 affinity than adult hemoglobins. A unique fetal hemoglobin (with no differences in red cell nucleosidetriphosphates) seems to be the basis of the higher O_2 affinity and the reduced Bohr effect of fetal blood in the viviparous blenny, *Zoarces viviparus,* (Weber and Hartvig 1984). Anatomical connections between mother and fetus are lacking in this species.

Different O_2 affinities can also be achieved by allosteric modulation of hemoglobin as well as by different concentrations of respiratory pigments. In the erythrocytes of mammals 2,3-bisphosphoglycerate reduces O_2 affinity of hemoglobin by binding to (and thus stabilizing) the deoxy form. In ectothermic vertebrates ATP is the main modulator of hemoglobin, often in combination with guanosine triphosphate (GTP) (which is usually more effective than ATP), other nucleosidetriphosphates, and inositolphosphates (for review, see Urich 1990; Weber and Jensen 1988). In the viviparous sea perch, *Embiotoca lateralis,* the O_2 affinity of fetal blood is higher, due to a unique fetal hemoglobin as well as decreased concentrations of nucleoside-triphosphate and hemoglobin in the fetal erythrocytes (Ingermann and Terwilliger 1984). Phosphate-free hemoglobin from new-born dogfish (*Squalus acanthias*) has a higher O_2 affinity than hemoglobin from the adult shark, suggesting the existence of a unique fetal hemoglobin.

A lower content of allosteric modulators in fetal than in maternal blood has also been reported of the caecilian *Typhlonectes* and the garter snake (see Table 1).

Hazards from Oxygen

Oxygen is indispensable for all higher forms of life, yet it is a mixed blessing because, at all concentrations, oxygen is highly toxic. When oxygen accumulated in the atmosphere, this was probably the worst ever global pollution of the environment. In order to survive, organisms had to develop defense mechanisms against oxygen toxicity.

The dualism of oxygen is also reflected in the way in which the therapeutic use of oxygen was discussed in the eighteenth and nineteenth centuries. While some physicians and scientists regarded oxygen as a cure for almost all evils, others were aware of the hazards from oxygen (for review, see Gilbert 1981). Among them was the influential chemist Justus von Liebig (1803–1873) from Gießen who considered food as an antioxidant that, by binding oxygen, counteracted the oxidizing power of the atmosphere. In his view, death from starvation was a consequence of the oxidation of body mass. Pasteur observed that some microorganisms were very sensitive to oxygen ("the air kills them"), and he proposed the term *anaerobes (anaérobies)* for this class of organisms (Pasteur 1863). Some 10 years later, Paul Bert showed that aerobes are also affected by oxygen toxicity.

The mechanism by which oxygen exerts its deleterious effects on living matter remained obscure for a long time. In the early 1950s Rebeca Gerschman noticed striking similarities between tissue damage caused by radiation and hyperbaric oxygen. She postulated that oxygen toxicity must be

Table 1. Some adaptations to maternal–embryonic gas exchange in viviparous ectothermic vertebrates

	Elasmobranchii (Chondrichthyes)	Osteichthyes (bony fish)	Amphibia	Reptilia
Morphological adaptations				
Placenta or placenta-like structures	Some sharks	Follicular placenta[a] Trophotaenial placenta (*Ameca splendens*)[c]		Chorioallantoic placenta[b] Yolk sac placenta[b]
Other structures	Trophonemata in skates and rays	Specialized epithelia of various fetal organs: fins, gut, pericard, yolk sac. Specialized ovarian structures[a]	"Branchial placenta" in apodans (external gills – vascularized uterus)[d]. Vascularized embryonic tail and uterus wall (*Nectophrynoides*, Anura)[e]	
Physiological adaptations in fetal blood				
Unique fetal hemoglobin with high O_2 affinity	Dogfish (*Squalus acanthias*)[f]	Blenny (*Zoarces viviparus*)[g] Sea perch (*Embiotoca lateralis*)[h]		
Lower content of allosteric modulators of hemoglobin in fetal blood cells	Nucleosidetriphosphates (*Squalus acanthias*)[i]	Sea perch (*Embiotoca lateralis*)[h]	Typhlonectes compressicaudus (Apoda)[j]	Garter snake (*Thamnophis elegans*)[k]

[a] Wourms (1981), [b] Yaron (1985), [c] Lombardi and Wourms (1985a, b), [d] Exbrayat and Delsol (1985), [e] Wake (1980), [f] Weber et al. (1983), [g] Weber and Hartvig (1984), [h] Ingermann and Terwilliger (1984), [i] Bartlett (1982), [j] Garlick et al. (1979), [k] Berner and Ingermann (1988)

brought about by the same mechanism as radiation damage: generation of free radicals (Gerschman et al. 1954).

Oxygen itself is not toxic because it reacts so slowly, but highly reactive compounds are formed if O_2 is only partially reduced, e.g., the superoxide anion $O_2^{-\cdot}$ (a free radical), hydrogen peroxide H_2O_2, and the hydroxyl radical $OH^{\cdot}$ (for review, see Newsholme and Leech 1983). These can damage macromolecules and cell membranes. Given the high toxicity of partially reduced oxygen, it appears sensible that, in energy metabolism, oxygen is involved only at the very end of a long sequence of anaerobic reactions (see also section on "Why aerobic metabolism?") in just one reaction where provisions have been made for handling it safely (for review, see Stryer 1988). On the other hand, the ingenuity of living systems can hardly be better demonstrated than by the capacity of certain cells that turn evil to good by using oxygen radicals to destroy foreign agents. Phagocytotic cells, for instance, have special enzymes that produce toxic oxygen species from O_2 (respiratory burst) in order to destroy engulfed particles such as pathogens (Badwey and Karnovsky 1980; Williams 1985). In chronic granulomatous disease, a defect in this mechanism makes patients highly susceptible to bacterial infections (usually fatal within the first decade of life).

Oxygen is essential for life, and yet all organisms encountering oxygen have had to develop strong antioxidant defense mechanisms. These mechanisms are not 100% safe. The effects of reactive oxygen species is regarded as one of the causes of irreversible changes in aging multicellular organisms (for review, Cutler 1984). This dilemma has led some authors to interpret, tongue in cheek, breathing as fatal addiction to oxygen: "The first inhalation (occurring at birth) is sufficient to make oxygen addiction permanent . . . which addiction invariably ends in death" (cf. Gilbert 1981). From what has been discussed in this chapter, one must face the fact that the situation is worse still: Man and all viviparous animals become addicted to oxygen long before parturition and are born as oxygen addicts.

References

Alberts B, Bray D, Lewis J, Raff M, Roberts K, Watson DJ (1989) Molecular biology of the cell, 2nd edn. Garland, New York

Amoroso EC (1968) The evolution of viviparity. Proc Roy Soc Med 61: 1188–1200

Badwey JA, Karnovsky ML (1980) Active oxygen species and the functions of phagocytic leukocytes. Annu Rev Biochem 49: 695–726

Balaban RS (1990) Regulation of oxidative phosphorylation in the mammalian cell. Am J Physiol 258 (Cell Physiol 27): C377–C389

Bartlett GR (1982) Phosphate compounds in red cells of two dogfish sharks: *Squalus acanthias* and *Mustellus canis*. Comp Biochem Physiol 73A: 135–140

Berner NJ, Ingermann RL (1988) Molecular basis of the difference in oxygen affinity between maternal and foetal red blood cells in the viviparous garter snake *Thamnophis elegans*. J Exp Biol 140: 437–453

Boyle R (1668) New experiments, concerning the relation between light and air. Philos Trans R Soc 2 (31): 581–600

Cutler RG (1984) Antioxidants, aging and longevity. In: Pryor WA (ed) Free radicals in biology, Vol. 6. Academic, New York, pp 371–428
Darnell J, Lodish H, Baltimore D (1986) Molecular cell biology. Scientific American Books, New York
Exbrayat J-M, Delsol M (1985) Reproduction and growth of *Typhlonectes compressicaudus* – a viviparous gymnophione. Copeia 4: 950–955
French TJ, Holness MJ, McLennan PA, Sugden MC (1988) Effects of nutritional status and acute variation in substrate supply on cardiac and skeletal-muscle fructose 2,6-bisphosphate concentrations. Biochem J 250: 773–779
Garlick RL, Davis BJ, Farmer M, Fyhn HJ, Fyhn UEH, Noble RW, Powers DA, Riggs A, Weber RE (1979) A fetal-maternal shift in the oxygen equilibrium of the hemoglobin from the viviparous caecilian, *Typhlonectes compressicauda*. Comp Biochem Physiol 62A: 239–244
Gerschman R, Gilbert DL, Nye SW, Dwyer P, Fenn WO (1954) Oxygen poisoning and X-irradiation: a mechanism in common. Science 119: 623–626
Gilbert DL (1981) Perspective on the history of oxygen and life. In: Gilbert DL (ed) Oxygen and living processes. Springer, New York Heidelberg Berlin, pp 1–43
Harold FM (1986) The vital force: a study of bioenergetics. Freeman, New York
Hochachka PW, Guppy M (1987) Metabolic arrest and the control of biological time. Harvard University Press, Cambridge
Hogarth PJ (1976) Viviparity. Arnold, London
Ingermann RL, Terwilliger RC (1984) Facilitation of maternal-fetal oxygen transfer in fishes: anatomical and molecular specializations. In: Seymour RS (ed) Respiration and metabolism of embryonic vertebrates. Junk, Dordrecht, pp 1–15
Jensen A, Hohmann M, Künzel W (1987) Dynamic changes in organ blood flow and oxygen consumption during acute asphyxia in fetal sheep. J Developm Physiol 9: 543–559
Kammer AE, Heinrich B (1978) Insect flight metabolism. Adv Insect Physiol 13: 133–228
Korsgaard B, Weber RE (1989) Maternal-fetal trophic and respiratory relationships in viviparous ectothermic vertebrates. Adv Comp Environ Physiol 5: 209–233
Krebs HA (1972) The Pasteur effect and the relations between respiration and fermentations. Essays Biochem 8: 1–34
Lagunas R (1981) Is Saccharomyces cerevisiae a typical facultative anaerobe? Trends Biochem Sci 6: 201–203
Lipmann F (1941) Metabolic generation and utilization of phosphate bond energy. In: Nord FF, Werkman CH (eds) Advances in enzymology and related subjects Vol 1. Interscience, New York, pp 99–162
Lombardi J, Wourms JP (1985a) The trophotaenial placenta of a viviparous goodeid fish. I. Ultrastructure of the internal ovarian epithelium, the maternal component. J Morphol 184: 277–292
Lombardi J, Wourms JP (1985b) The trophotaenial placenta of a viviparous fish II. Ultrastructure of the trophotaenial placenta, the embryonic component. J Morphol 184: 293–309
Mitchell P (1961) Coupling of phosphorylation to electron and hydrogen transfer by a chemi-osmotic type of mechanism. Nature 191: 144–148
Nelson N, Taiz L (1989) The evolution of H^+-ATPases. Trends Biochem Sci 14: 113–116
Newsholme EA, Leech AR (1983) Biochemistry for the medical Sciences. Wiley, Chichester
Passonneau JV, Lowry OH (1962) Phosphofructokinase and the Pasteur effect. Biochem Biophys Res Commun 7: 10–15
Pasteur L (1861) Expériences et vues nouvelles sur la nature de fermentations. C R Acad Sci 52: 1260–1264
Pasteur L (1863) Recherches sur la putrefaction. Comp Rend Acad Sci 56: 1189–1194
Racker E (1980) From Pasteur to Mitchell: a hundred years of bioenergetics. Fed Proc 39: 210–215

Rutherford AW (1989) Photosystem II, the water-splitting enzyme. Trends Biochem Sci 14: 227–232
Seymour RS (1984) Respiration and metabolism of embryonic vertebrates. Junk, Dordrecht
Siesjö BK (1978) Brain energy metabolism. Wiley, Chichester
Storey KB (1985) A re-evaluation of the Pasteur effect: new mechanisms in anaerobic metabolism. Mol Physiol 8: 439–461
Stryer L (1988) Biochemistry. Freeman, New York
Urich K (1990) Vergleichende Biochemie der Tiere. Fischer, Stuttgart
Van Schaftingen E (1987) Fructose 2,6-bisphosphate. Adv Enzymol Relat Areas Mol Biol 59: 315–395
Van Schaftingen E, Hue L, Hers HG (1980) Fructose 2,6-bisphosphate, the probable structure of the glucose- and glucagon-sensitive stimulator of phosphofructokinase. Biochem J 192: 897–901
Wake MH (1980) The reproductive biology of *Nectophrynoides malcolmi* (Amphibia: Bufonidae), with comments on the evolution of reproductive modes in the genus *Nectophrynoides*. Copeia 2: 193–209
Wald G (1966) On the nature of cellular respiration. In: Kaplan NO, Kennedy EP (eds) Current aspects of biochemical energetics. Academic, New York, pp 27–32
Weber RE, Hartvig M (1984) Specific fetal hemoglobin underlies the fetal-maternal shift in blood oxygen affinity in a viviparous teleost. Mol Physiol 6: 27–32
Weber RE, Jensen FB (1988) Functional adaptations in hemoglobins from ectothermic vertebrates. Annu Rev Physiol 50: 161–179
Weber RE, Wells RMG, Rosetti JE (1983) Allosteric interactions governing oxygen equilibria in the hemoglobin system of the spiny dogfish, *Squalus acanthias*. J Exp Biol 103: 109–120
Wegener G (1981) Comparative aspects of energy metabolism in nonmammalian brains under normoxic and hypoxic conditions. In: Stefanovich V (ed) Animal models and hypoxia. Pergamon, Oxford, pp 87–109
Wegener G (1988) Oxygen availability, energy metabolism, and metabolic rate in invertebrates and vertebrates. In: Acker H (ed) Oxygen sensing in tissues. Springer, Berlin Heidelberg New York Tokyo, pp 13–35
Wegener G (1990) Elite invertebrate athletes: flight in insects, its metabolic requirements and regulation and its effects on life span. In: Nazar K, Kaciuba-Uscilko H, Terjung RL, Budohoski L (eds) International perspectives in exercise physiology. Human Kinetics, Champaign, pp 83–87
Wegener G, Michel R, Newsholme EA (1986a) Fructose 2,6-bisphosphate as a signal for changing from sugar to lipid oxidation during flight in locusts. FEBS Lett 201: 129–132
Wegener G, Michel R, Thuy M (1986b) Anoxia in lower vertebrates and insects: effects on brain and other organs. Zool Beitr 30: 103–124
Wegener G, Beinhauer I, Klee A, Newsholme EA (1987) Properties of locust muscle 6-phosphofructokinase and their importance in the regulation of glycolytic flux during prolonged flight. J Comp Physiol B 157: 315–326
Wegener G, Krause U, Thuy M (1990) Fructose 2,6-bisphosphate and glycolytic flux in skeletal muscle of swimming frog. FEBS Lett 267: 257–260
Wegener G, Bolas NM, Thomas AAG (1991) Locust flight metabolism studied in vivo by ^{31}P NMR spectroscopy. J Comp Physiol B 161: 247–256
Williams RJP (1985) The necessary and desirable production of radicals in biology. Philos Trans R Soc Lond B 311: 593–604
Wourms JP (1981) Viviparity: the maternal-fetal relationship in fishes. Am Zool 21: 473–515
Yaron Z (1985) Reptilian placentation and gestation: structure, function and endocrine control. In: Gans C (ed) Biology of the reptilia, vol 15 B. Wiley, New York, pp 527–604

Maternal Vascular Adaptation to Improve Oxygen Supply

W. MOLL, A. NIENARTOWICZ, R. BAUSTÄDTER, H. PARISH, and A. RUIDER

Fetal arterial oxygen saturation, which may be considered a reliable indicator of fetal oxygenation, is controlled by the maternal placental blood flow. This can be theoretically derived with any degree of precision and complexity. Fetal oxygen consumption, diffusion capacity of the placental membrane, umbilical blood flow, and oxygen affinity of maternal and fetal blood are important parameters of this relationship. In Fig. 1A an experimental set-up for the guinea pig is shown which allows the maternal placental blood flow to be adjusted to different values, and permits experimental study of the effect of blood flow variations on fetal arterial saturation (Herberger and Moll 1976). Using this experimental approach, one can demonstrate the control of fetal arterial saturation by the maternal placental blood flow (Fig. 1B). It can be seen that fetal arterial O_2 saturation is strongly dependent on maternal placental flow per fetal weight in the range of low blood flow rates. This curve illustrates how fetal arterial O_2 saturation is controlled by maternal placental flow during labor when maternal blood flow is acutely reduced; this is possibly also true during some forms of intrauterine growth retardation when maternal placental flow is chronically reduced.

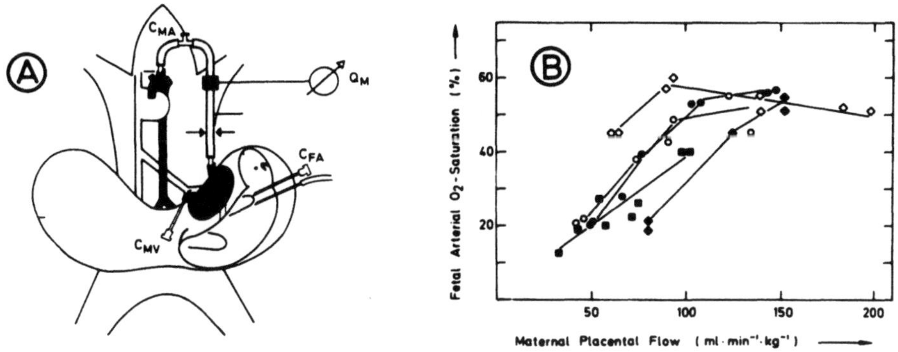

Fig. 1A, B. A Experimental arrangement to study the dependence of the fetal arterial O_2 saturation related to maternal placental blood flow (Herberger and Moll 1976). **B** Oxygen saturation related to maternal placental flow in the guinea pig (Moll, Klappstein, and Bartag, unpublished data)

Maternal Placental Flow Rate is Adapted to Fetal Growth

During pregnancy, fetal oxygen saturation is kept constant despite a continuous increase in fetal weight and oxygen consumption. This may indicate that fetal arterial oxygenation controls the maternal placental flow. As usually occurs in living systems, by virtue of biological control systems the physically controlled factor seems to become the determining factor; sense replaces cause as pertinent factor. During pregnancy the placental blood flow rises in parallel with fetal weight. This is true for the sheep, where the fetus grows rather slowly, as well as for the rat where fetal weight doubles every 2 days (Fig. 2). In this way, the maternal placental flow per fetal weight and thus fetal arterial oxygen saturation remain fairly constant. The fact that fetal arterial oxygen saturation is kept constant during pregnancy was the first important finding obtained in the chronic cannulated sheep preparation by Meschia et al. (1965). The adaptation of the maternal placental flow to fetal growth is based on two events: 1. the establishment of the placental vascular bed and 2. the widening of the uterine arteries which supply the placenta.

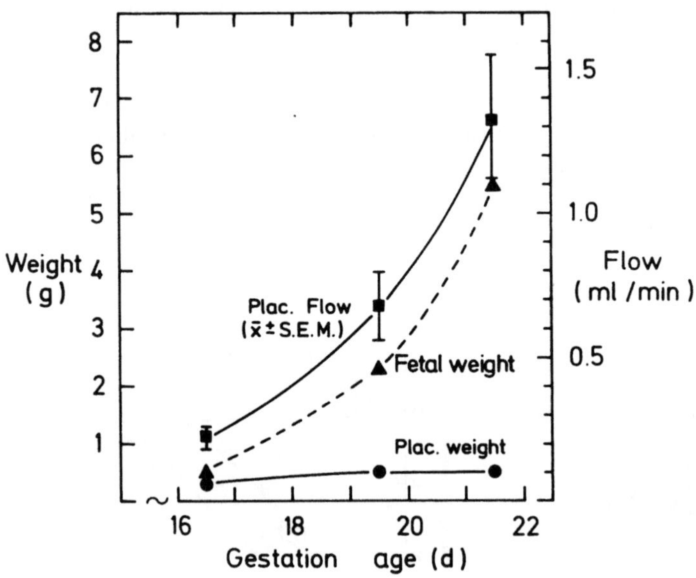

Fig. 2. Maternal placental flow and fetal weight in the rat (Parish and Moll, unpublished data)

Maternal Placental Flow Rises Due to Enlargement of the Maternal Placental Bed

The placental vascular bed has a high level of vascular conduction. When the local microvascular system with low conduction is replaced by the maternal placental vascular bed, a major increase in blood flow results. There are two different ways of establishing the maternal placental vascular bed as indicated in Fig. 3A. The epitheliochorial placenta (Fig. 3A) secretes angiogenic factors which induce proliferation of maternal endothelial cells and enlargement of the endometrial microvascular system (Reynolds et al. 1987; Reynolds and Redmer 1988). However, a clear concept of the cascade of signals initiating the formation of the maternal placental vascular bed is still lacking. In the hemochorial placenta, the maternal vascular bed is formed by the trophoblast itself (Fig. 3A). The trophoblast constructs a system of channels which are connected to the maternal arteries and veins. The maternal vascular bed in the hemochorial placenta has a high level of conduction (Fig. 3B). The normal maternal placental flow (around 10 ml/min in the guinea pig near term) requires only a few mmHg driving pressure difference. Maternal placental flow is controlled by the placental arteries under these conditions.

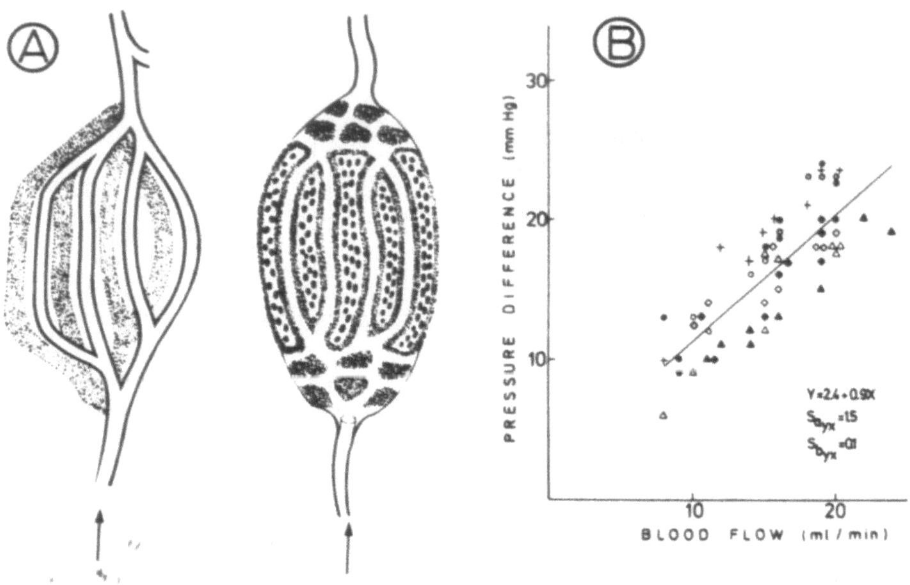

Fig. 3A, B. A Formation of the maternal placental vascular bed by maternal tissue (*right*, epitheliochorial placenta) and by the trophoblast (*left*, hemochorial placenta). **B** Driving pressure difference across the maternal placental vascular bed related to the blood flow rate in the guinea pig. The graph demonstrates high conductance of the maternal placental vascular bed (Herberger and Moll 1976)

Kurtz et al. (1986) have shown that hypoxia stimulates the secretion of growth factors affecting erythropoiesis. One may speculate that the development of the maternal placental vascular bed is controlled by oxygen-dependent stimulation of secretion of growth factors. However, there are only few hints that a direct coupling exists between fetal oxygenation and the formation of the maternal placental vascular bed. Thus, Bacon et al. (1984) showed in the guinea pig that placental vasculature increases in size during hypoxia.

Centripetal Dilation of the Placental Arteries is a Second Mechanism of Blood Flow Adjustment

During gestation, the uterine arteries supplying a placenta show an extensive increase in diameter (Fig. 4A). The final portion of a mesometrial artery in the guinea pig increases in diameter sevenfold (Fig. 4B). Widening starts at the very end of the artery. The final portion of the arteries shows a stepwise increase in diameter in the area where it comes close to the uterus. For the sake of further blood flow adaptation to fetal growth, local peripheral resistance is virtually abolished (Moll et al. 1983).

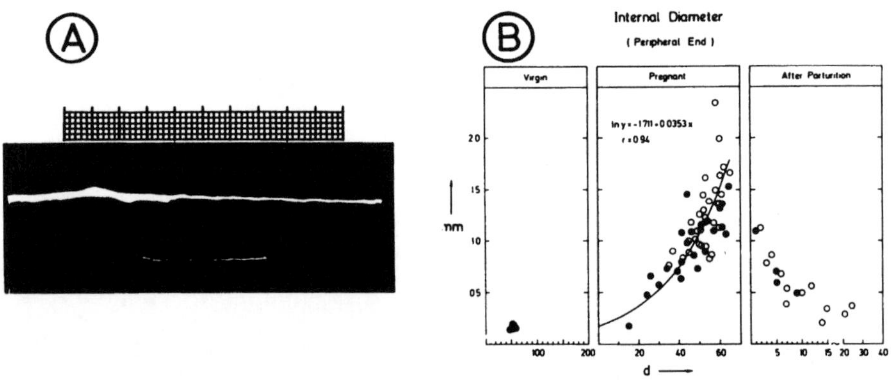

Fig. 4A, B. A Photogram of a mesometrial artery from a nonpregnant and a pregnant guinea pig. **B** Increase in diameter of a mesometrial artery during pregnancy (Moll et al. 1983)

Arterial Dilation is Based on Remodelling of the Arterial Wall

The extensive local dilation is not based on smooth muscle relaxation. The differences in diameter are also seen on arteries of pregnant and nonpregnant guinea pigs after they have been equally relaxed by papaverine (Moll and

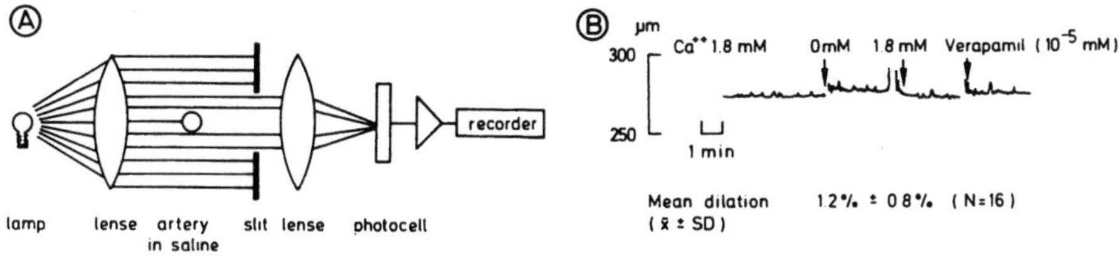

Fig. 5A, B. A Diagram showing an optical set-up for continuous registration of the external diameter of an artery. **B** External diameter of a mesometrial artery after application of calcium-free Tyrode's solution (Baustädter 1987)

Götz 1985). In order to measure the dilation induced by smooth muscle relaxation, uterine arteries were removed from a nonpregnant animal and placed in an optical device, where the diameter could be measured continuously (Fig. 5A). The arterial diameter increased by only about 1% (Baustädter 1987; Fig. 5B) when they were submerged in a calcium-free salt solution, while during pregnancy the diameter increases sevenfold.

Experiments in guinea pigs show that the widening process is caused by extensive remodeling of the arterial wall during pregnancy. The final portion of the mesometrial artery supplying a guinea pig placenta increases 50-fold in weight. [^{3}H]thymidine incorporation increases 40-fold (Makinoda and Moll 1986), and DNA mass 30 times (Moll et al. 1983). Collagen synthesis is increased. When the arteries are excised and incubated in [^{3}H]proline-containing solutions, one can see that the specific activity of hydroxyproline rises much faster in arteries from animals in midpregnancy than in nonpregnant animals (Ruider 1988; Fig. 6A). At the same time, the release of hydroxyproline into the incubation solution is higher in pregnant than in nonpregnant animals (Fig. 6B). However, it seems that part of the liberated hydroxyproline is derived from newly synthesized collagen rather than from pre-existing collagen. Thus, the question of the mechanism of arterial widening is not yet finally answered.

Neither the control mechanism of arterial widening, nor the coupling with fetal oxygenation has been clarified so far. We have some hints that estradiol may be involved – the initial stage of arterial widening and growth can be induced by long-lasting estradiol treatment (Moll et al. 1988) – but we are far from completely understanding it.

According to the constant shear stress theory (Rodbard 1975), arteries widen whenever the shear stress caused by the flowing blood is increased. According to this theory, the widening of the uterine arteries supplying a placenta is caused by the increased blood flow following the formation of the placental vascular bed. However, it can be shown that increased shear stress is not a necessary stimulus of gestational dilation. In the rat, the blood flow rate across the uterine artery can be reduced by ligating the uterine artery. Under these conditions the uterine blood supply is provided exclusively by the

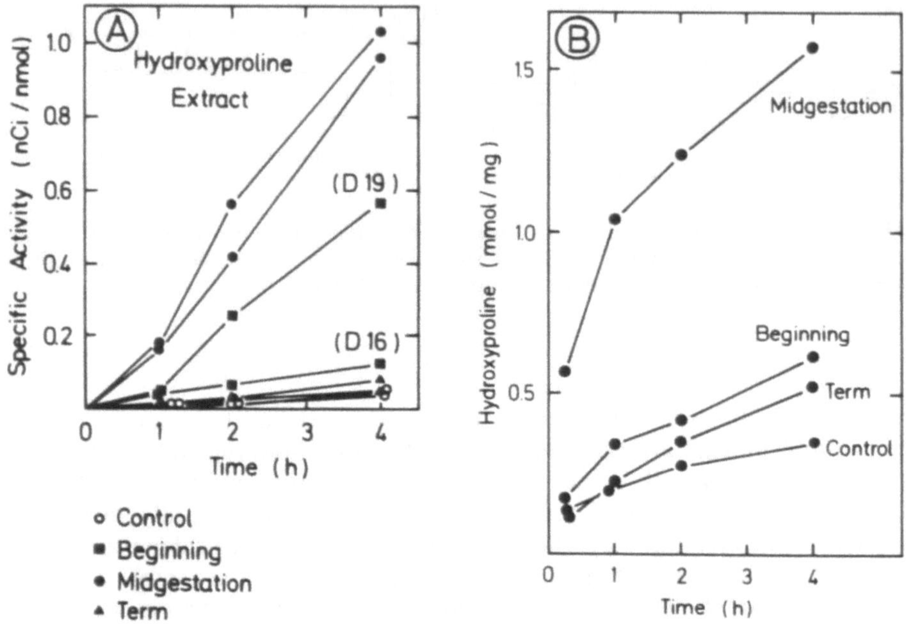

Fig. 6 A, B. **A** Specific activity of hydroxyproline in mesometrial arteries during incubation in [^{3}H]proline. **B** Liberation of hydroxyproline from mesometrial arteries during incubation (Ruider 1988)

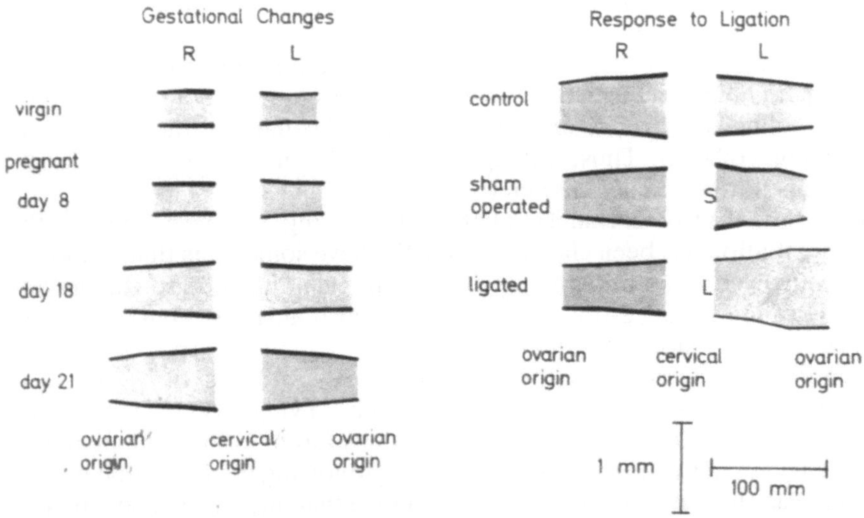

Fig. 7. Schematic diagram showing the adaptive changes in the uterine arcade of the rat during normal pregnancy and after ligation of the left uterine artery (Nienartowicz et al. 1989)

ovarian artery. Even after the ligation, normal widening of the uterine artery was seen (Moll et al. 1988; Fig. 7), demonstrating that widening proceeds even when shear stress is decreased. Shear stress, however, may well induce an emergency mechanism to provide sufficient blood supply even when the normal mechanism of increasing blood supply through the uterine artery fails. When the uterine artery is ligated in the rat, the ovarian artery enlarges to such an extent that the overall vascular conductance of the uterine arterial system is restored to the value in the intact animal.

Conclusions

While biophysically the maternal placental flow controls fetal arterial oxygen saturation, placental blood flow during pregnancy behaves as if it is regulated by fetal oxygen saturation. The basis of the adjustment of the maternal placental flow rate is the formation of the maternal placental vascular bed and the peripheral widening of the supplying arteries by tissue growth and remodeling of the arterial wall. The clarification of the basic mechanisms and the control systems that couple directly or indirectly maternal placental blood flow with fetal oxygenation remains a task for the future.

References

Bacon BJ, Gilbert RD, Kaufmann P, Smith AD, Trevino FT, Longo LD (1984) Placental anatomy and diffusing capacity in guinea pig following long-term maternal hypoxia. Placenta 5: 475–488

Baustädter R (1987) Akute Dilatation und Konstriktion präuteriner Mesometrialarterien des Meerschweinchens in vitro. Thesis, University of Regensburg

Herberger J, Moll W (1976) The flow resistance of the maternal placental vascular bed of anesthetized guinea pigs. Z Geburtshilfe Perinatol 180: 61–66

Kurtz A, Jelkmann W, Pfuhl A, Malmstroem K, Bauer C (1986) Erythropoietin production by fetal mouse liver cells in response to hypoxia and adenylate cyclase stimulation. Endocrinology 118: 567–572

Makinoda S, Moll W (1986) Desoxyribonucleic acid synthesis in mesometrial arteries of guinea pigs during oestrous cycle, pregnancy and treatment with oestradiol benzoate. Placenta 7: 189–198

Meschia G, Cotter J, Breathnach C, Barron D (1965) The hemoglobin, oxygen, carbon dioxide and hydrogen ion concentrations in the umbilical bloods of sheep and goats as sampled via indwelling plastic catheters. Q J Exp Physiol 50: 185–195

Moll W, Götz R (1985) Pressure-diameter curves of mesometrial arteries of guinea pigs demonstrate a non-muscular, oestrogen-inducible mechanism of lumen regulation. Pflugers Arch 404: 332–336

Moll W, Espach A, Wrobel KH (1983) Growth of mesometrial arteries in guinea pigs during pregnancy. Placenta 4: 111–124

Moll W, Nienartowicz A, Hees H, Wrobel KH, Lenz A (1988) Blood flow regulation in the uteroplacental arteries. Trophoblast Res 3: 83–96

Nienartowicz A, Link S, Moll W (1989) Adaptation of the uterine arcade in rats to pregnancy. J Dev Physiol 12: 101–108

Reynolds LP, Redmer DA (1988) Secretion of angiogenic activity by placental tissues of cows at several stages of gestation. J Reprod Fertil 83: 497–502

Reynolds LP, Millaway DS, Kirsch JD, Infeld JE, Redmer DA (1987) Angiogenic activity of placental tissues of cows. J Reprod Fertil 81: 233–240
Rodbard S (1975) Vascular caliber. Cardiology 60: 4–49
Ruider A (1988) Untersuchungen zum Kollagenumsatz in uterinen mesometrialen Arterien des Meerschweinchen während der Gravidität und nach Oestradiolbehandlung. Thesis, University of Regensburg

Oxygen Deficiency During Fetal Development and Delivery – Speculations About Clinical Conditions

W. Künzel, M. Hohmann, and G. Braems

There is no doubt that fetal development is dependent on a sufficient oxygen supply. This includes ovulation, the passage through the fallopian tube, implantation, and placentation itself. The establishment of the vascular system facilitates the oxygen transfer from the mother to the fetus and ultimately the oxygen transport to the fetal tissue (Bartels and Baumann 1972; Longo et al. 1972; Moll 1972; Wulf et al. 1972).

At term the fetus is liberated from its intrauterine environment, since the oxygen supply might be restricted past term. In addition, fetal weight gain is not paralleled by placental growth and maturation since there are limitations in the increase of the surface area and the fall of diffusion distances (Kaufmann 1981). Therefore, labor starts in normal pregnancies at a time when the uterine blood flow is still sufficient to compensate the reduction of uterine blood flow during labor.

Several clinical situations have to be considered in which this process is disturbed from ovulation to delivery (Künzel 1989). Fetal oxygen supply might be impaired in particular clinical situations such as intrauterine growth retardation, which is often a feature of twin and multiple pregnancies, intrauterine death, diabetes in pregnancy, and pregnancy-induced hypertension.

During the past years, many investigators have focused their interest on factors affecting the gas transfer across the placenta and the oxygen supply to the fetus (for review, see Carter 1989; Acheson et al. 1957). For clinical management during pregnancy and delivery, it is very important to understand the influence of the reduced umbilical and uterine blood flow on fetal oxygenation, to evaluate the fetal cardiovascular response to a reduced O_2 supply and to estimate the impact of mortality and morbidity on the fetus.

The aim of this chapter is to point out particular fetal hazards occurring during gestation which might be due to a diminished O_2 supply to the fetus. For that reason, facts and speculations concerning the O_2 delivery to the fetus presented in this chapter could be the basis for further investigations.

Table 1. Incidence of perinatal mortality: a comparison of perinatal survey from Hesse, FRG (1988), and the British Perinatal Mortality Survey 1958 (Gruenwald 1969)

Time of death	Hesse, FRG, 1988 *n*	%	Gruenwald 1969 *n*	%
Ante partum	152	52.2	2311	31.5
Intra partum	11	3.8	2270	30.9
First week after birth	128	43.9	2749	37.5
Total	291	0.6	7330	1.25

Ante Partum Fetal Death – The Problem of Fetal Surveillance

Perinatal mortality has decreased during the past few years tremendously. This mortality rate lies between five and seven deaths per thousand births in European countries (Steiner 1989). This figure represents half of that reported by Gruenwald in 1969, when referring to data from the second Report of the British Perinatal Mortality Survey in 1958 (Table 1), in which the perinatal mortality was 1.25%. This report shows that the perinatal mortality was equally distributed between ante partum death (31.5%), death during labor (30.9%), and death during the first week of life (37.5%). However, death during parturition is nowadays a rare occurrence often due to unpredictable complications during labor, i.e., cord prolapse and heavy bleeding. This accounts for about 3.8% of all perinatal deaths (Künzel 1989). Electronic fetal heart rate monitoring has reduced intra partum death and even morbidity despite hypoxic episodes during contractions (Review in Künzel and Hohmann 1989). Ante partum death, often related to insufficient fetal growth, is still a problem that has to be solved in the future. Table 2 shows ante partum fetal mortality in relation to fetal weight. Fetal growth retardation is a sign of insufficient fetal adaptation throughout pregnancy (Clapp et al. 1980).

In cases of ante partum fetal death during weeks 39–41 of gestation, the fetal weight is significantly lower (19%) than in the total population of the same gestational age (1.4%). There is also a strong relationship in this population between the incidence of ante partum mortality and birth weight: 14.3% of ante partum death occurred in the low birth weight group of 1000–1499 g and only 0.1% occurred in fetuses with a birth weight of 2500 g and more.

The relationship between ante partum death and fetal weight is even stronger in the group at less than 39 weeks of pregnancy. Fetal death occurred in 72.9% if the weight was 2500 g and less compared with an overall rate of only 29.1%. However, the higher mortality rate in this group is also combined with lower weight (Table 2). Similar results have been previously reported by Hughes et al. (1987).

Table 2. Ante partum fetal mortality in relation to fetal weight and compared with the total population of the Hesse perinatal survey 1986–1989

	Up to week 38 of gestation				Between weeks 39 and 41 of gestation			
	Ante partum fetal death (n = 414)		Total population (n = 32567)		Ante partum fetal death (n = 170)		Total population (n = 140815)	
Birth weight (g)	n	%	n	%	n	%	n	%
< 1000	20	4.83	546	1.68	0	0	7	0.005
1000–1499	99	23.91	1217	3.74	2	1.18	14	0.01
1500–1999	94	22.71	2250	6.91	7	4.12	113	0.08
2000–2499	89	21.50	5459	16.76	22	12.94	1798	1.28
≥ 2499	112	27.05	23095	70.91	139	81.76	138883	98.62
Total	414	100	32567	100	170	100	140815	100

The impact of a low oxygen condition on mortality and growth is also evident in populations living at a high altitude. Hematologic evidence of fetal hypoxia has been reported among newborn infants living at a high altitude in Bolivia (Ballew and Haas 1986). The high altitude newborn infant displays evidence of enhanced erythropoiesis and shows a reduced birth weight compared with the same ethnic group at a low altitude.

Indications for Fetal Oxygen Deficiency During Pregnancy

A fetus with normal growth and weight gain during pregnancy probably has a normal supply of oxygen and nutrients. Fetal weight gain can be easily assessed by measuring the symphysis-fundus distance and by ultrasonographic measurements of the biparietal diameter and the abdominal circumference of the fetus. Deviations from the normal growth curve are often associated with conditions such as pregnancy-induced hypertension (PIH), maternal hypotensive disease (Hohmann and Künzel 1990), severe diabetes in pregnancy, often combined with vascular complications (Lang and Künzel 1989), or maternal anemia (Paulone et al. 1987). The oxygen transfer to the fetus is assumed to be disturbed by the reduction of the oxygen transport capacity, i.e., the blood flow to the uterus and to the fetus and the oxygen-carrying capacity (Block et al. 1984; Clapp 1978; Clapp et al. 1980; Owens et al. 1987).

The following clinical disorders which are associated with a decreased O_2 supply during pregnancy will be analyzed in more detail.

Pregnancy-Induced Hypertension

PIH, predominantly present in younger and older women during pregnancy, is often associated with intrauterine growth retardation and fetal heart rate alterations during labor. In a group of 285 cesarean sections performed at the Department of Obstetrics and Gynecology in Gießen between 1985 and 1986 (Figs. 1, 2) the PIH rate with preterm fetuses and below the 10th percentile was 29% compared with 10% in normal pregnancies. The lighter babies had also a higher incidence of pathological fetal heart rate patterns, i.e., decelerations in relation to uterine contractions (69% vs 41%). At term, the same observations could be made, demonstrating clearly that fetal oxygenation was at the border of decompensation in the growth-retarded babies.

Diabetes in Pregnancy

The fetus in less controlled diabetic patients is usually macrosomic. It often dies in utero if the fetal weight is far more than the 90th percentile. In these cases, the placental mass per fetal weight is similar to that normal weight

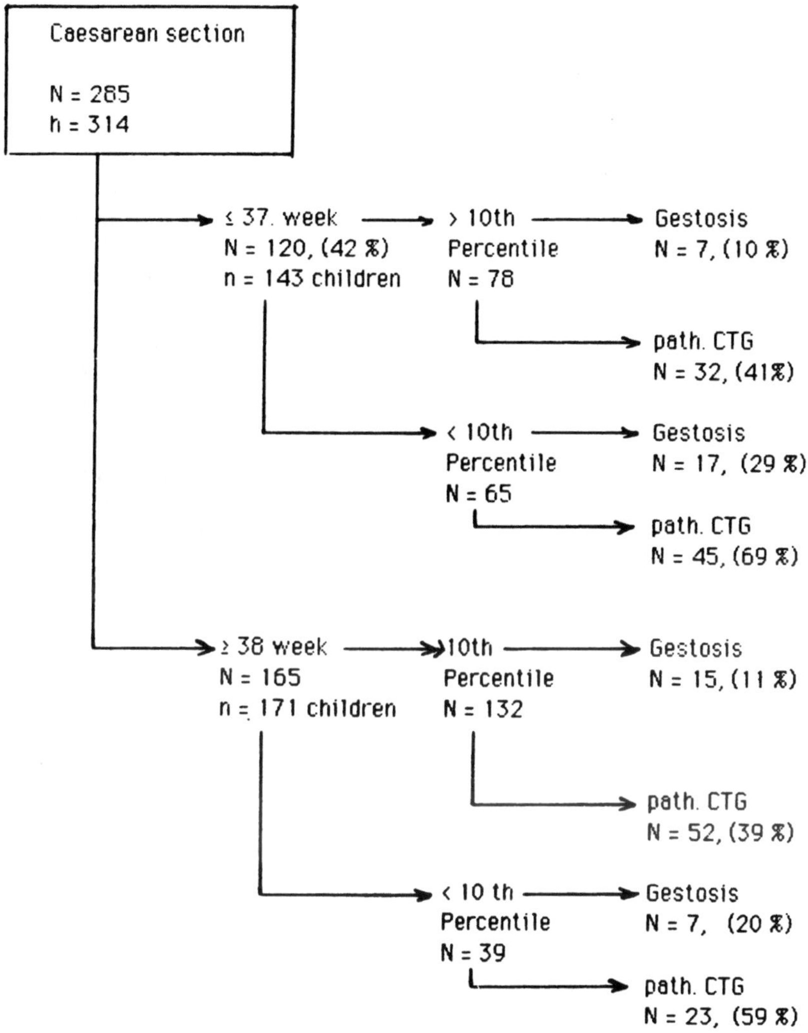

Fig. 1. Incidence of gestosis and pathological alterations in the cardiotocograms of 285 pregnancies with 314 children delivered by cesarean section at the Department of Obstetrics and Gynecology at the University of Giessen between 1985 and 1986. The group was divided into deliveries from week 38 and until week 37, and into growth-retarded and normal infants (< 10th percentile)

babies. Fetal O_2 supply is, however, disturbed by the increased diffusion distance in the fetal villi: the capillary surface area is reduced in relation to the total area of the villi, i.e., the diffusion distances are elevated (Emmrich 1981).

In poorly controlled diabetic patients with vascular diseases, the fetus is, however, despite a sufficient glucose transfer, not macrosomic, but reduced in weight (Fig. 3) (Lang and Künzel 1989). It may be speculated that the lower

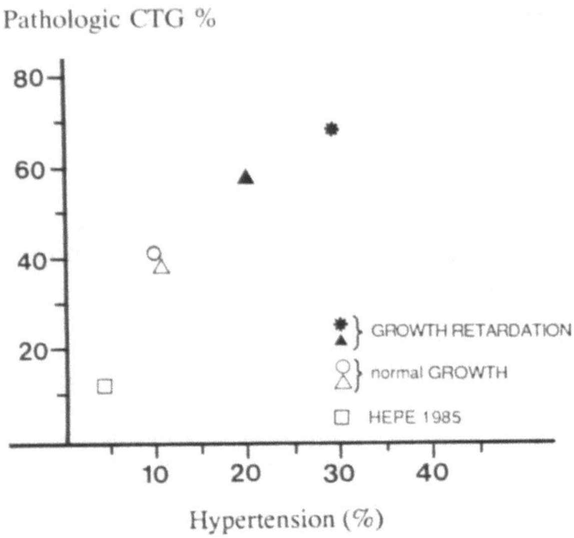

Fig. 2. The relationship between the incidence of decelerations in fetal heart rate and hypertension in fetuses with growth retardation and normal growth. The delivery occurred by cesarean section. In comparison, the data of the Hesse Perinatal Survey, 1985 *(HEPE 1985)* are shown. Fetal heart rate deterioration is associated with a higher incidence of hypertension in pregnancy

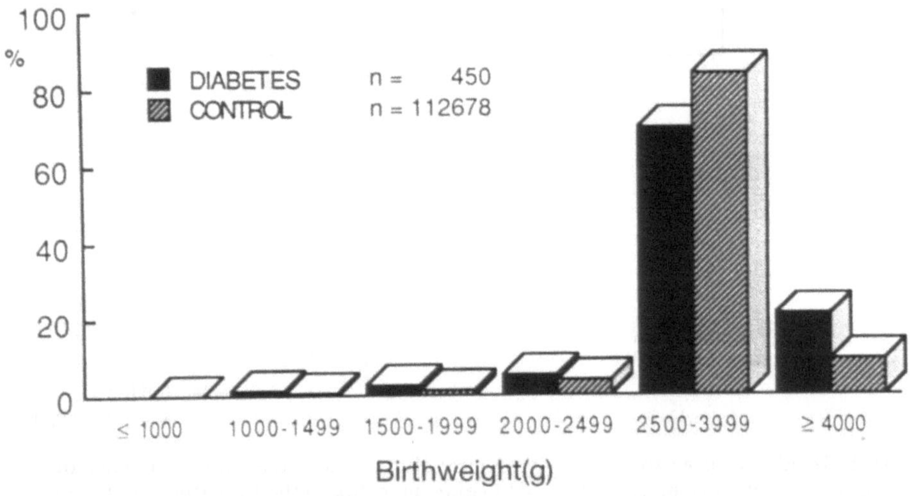

Fig. 3. Distribution of birth weight in children of diabetic mothers *(black shading)* compared with children of nondiabetic mothers *(control; lined shading)*. The percentage of newborns below 2500 g is increased in children with diabetic mothers (9% vs 6.3%; $p < 0.001$). The same applies to birth weight above 3999 g (21.2% vs 9.4%; $p < 0.001$). (From Lang and Künzel 1989)

uterine blood flow caused by vascular alterations limits the oxygen transfer and the incorporation of glucose into the fetal cells as an oxygen-dependent process. This speculation, however, needs to be proved in further investigations.

Multiple Pregnancy

Impairment in fetus growth is influenced by the number of fetuses in one uterus. This is well-known in human fetuses (for review, see Dawes 1968) and also valid for various other species (Ibsen 1928; Wigglesworth 1964). Less attention has been given, however, to the fact that multiple pregnancies show not only fetal growth retardation, but in 50% of them there is also a termination of pregnancy in the 36th week of gestation or earlier, and not in the 40th week as is normal for single fetuses (Fig. 4). Multiple fetus pregnancies which end between weeks 37 and 39 show a higher incidence (35.3%) of small for gestational age babies (2000 g and less) than single fetuses (5%) (Table 3). These risks are often combined with premature labor, cervical incompetence, and softening with severe retardation. Decelerations in fetal heart rate are common (Fig. 5) and indicate fetal hypoxemia and fetal deterioration. In this context it may be speculated that "mild" fetal hypoxia in addition to growth retardation and early labor is the trigger signal for a biochemical process which induces labor.

Fetal Supervision During Pregnancy and Labor

It has been shown in animal experiments that uterine contractions are associated with a reduction in uterine blood flow (Harbert 1982; Greiss 1965; Moll and Künzel 1974). This decrease in flow is related to the strength and

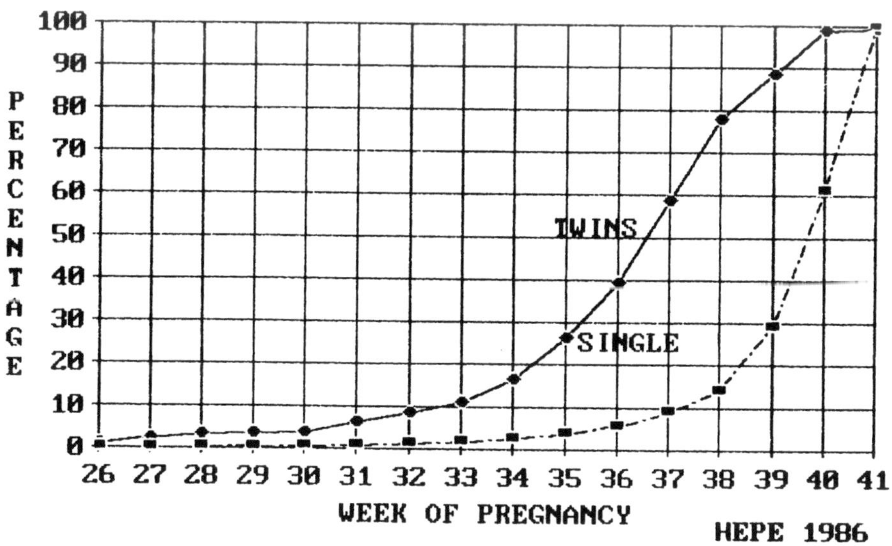

Fig. 4. The termination of pregnancy in single fetuses and in twins. Fifty percent of twin pregnancies are terminated between the 36th and 37th week of gestation (Hesse Perinatal Survey 1986, *HEPE 1986*)

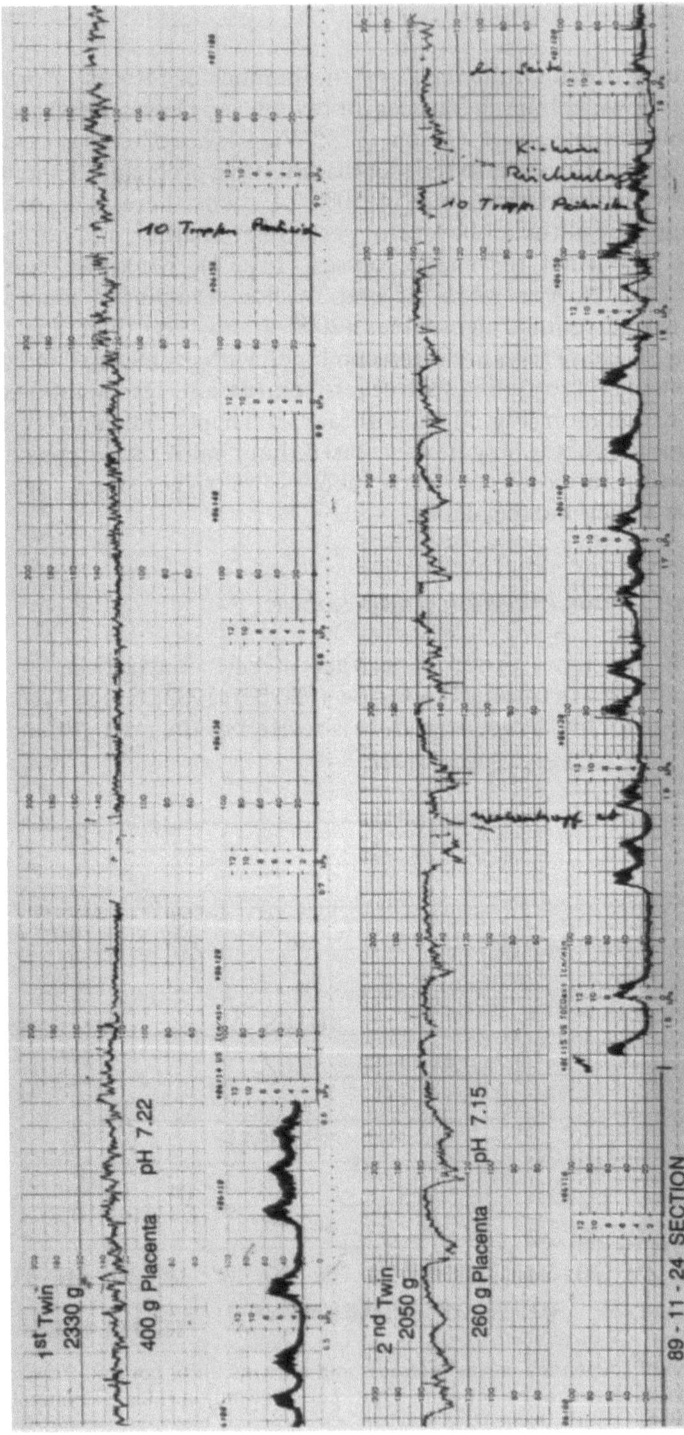

Fig. 5. Fetal heart rate of twins in the 38th week of pregnancy. Both fetuses were growth retarded: 2330 g and 2050 g. The second twin exhibited deceleration of fetal heart rate (FHR). The placental mass of the first twin was 172 g/kg fetus (60.6%) and of the second twin 126 g/kg fetus (39.4%). Hypoxia was evident by the decelerations of FHR and by the pH in the umbilical artery (7.22 vs 7.15). The twins were delivered by caesarean section

Table 3. Comparison of birth weights in twins and single fetuses born between weeks 37 and 39 of gestation

Weight (g)	Weeks 37–39 of pregnancy	
	Twins (n = 338D)[a] (%)	Single fetuses (n = 8345)[b] (%)
≤ 1000	0.3	0.0
1000–1499	1.2	0.0
1500–1999	3.6	0.3
2000–2499	30.2	4.7
2500–3999	64.8	91.5
> 3999	0.0	3.5

[a] From Hesse Perinatal Survey 1986
[b] From Hesse Perinatal Survey 1987

duration of contraction. Normal contractions during labor, i.e., not more than 50 mmHg and 30–45 s duration, reduce the uterine blood flow by about 50% (Künzel 1973). This does not harm fetal oxygen consumption because of the high "buffer capacity" of the uterine and umbilical circulation (Fig. 6) (Künzel and Moll 1972; Künzel et al. 1972, 1974, 1977).

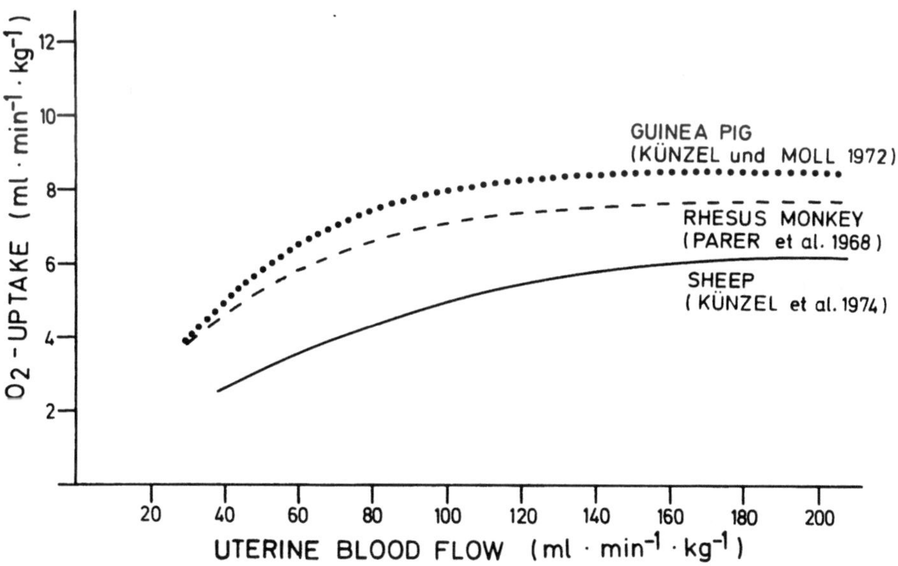

Fig. 6. The relationship between uterine oxygen consumption and uterine blood flow in guinea pigs (Künzel and Moll 1972), rhesus monkey (Parker et al. 1968), and sheep (Künzel et al. 1974), in acute experiments with animals under anesthesia. Oxygen consumption fell if a critical limit of uterine blood flow was reached. Chronic experiments have confirmed a similar relationship of flow and O_2 consumption. It is of importance that the same relationship could be established for animals of different size and placental structure

In pathologic conditions, however, the uterine and umbilical blood flows are assumed to be low, and contractions reduce the flow below a critical limit where fetal oxygen consumption falls. This was confirmed by Clapp (1978), Itzkowitz et al. (1983, 1987), and recently by Edelstone et a. (1985) for the umbilical circulation. The fetal response to this diminished O_2 supply is a deceleration of the heart rate (Junge et al. 1977; Künzel et al. 1983; see review, Künzel 1986). These observations have formed the theoretical basis for the use of the contraction stress test for years (Hohmann et al. 1986). Clinical observations have demonstrated the contraction stress test to be a valuable tool in the early detection of hazards to the fetus. In cases of reduced uterine blood flow, the heart rate decelerations occur before the oscillations and accelerations of the fetal heart rate attenuate, i.e., before the fetus shows signs of a redistribution of its blood flow to vital organs. The evaluation of 103 cardiotocograms of growth-retarded babies delivered by cesarean section showed that 61 had decelerations following an oxytocin challenge test (Table 4). In 42 infants, no decelerations were observed. In the latter group, cesarean section was performed because of obstruction of labor and other reasons. The first group had a higher incidence of acceleration and oscillation

Table 4. Analysis of 103 cardiotocograms of grwoth retarded babies delivered by primary and secondary cesarean sections

	Primary cesarean section (n = 61) (%)	Secondary cesarean section (n = 42) (%)
Decelerations		
No	0	88
Questionable	0	12
Yes	100	0
Baseline fetal heart rate (b/min)		
≤ 140	18	36
141–150	39	17
≥ 151	43	48
Accelerations (per 30 min)		
≥ 5	26	86
4–1	21	7
0	53	7
Oscillations of amplitude (b/min)		
≥ 10	20	74
5–10	28	19
≤	53	7
Long-term irregularities (per min)		
≥ 6	51	90
2–5	26	7
≤ 1	23	2

loss, indicating redistribution in the cardiovascular system with preference to the heart and brain of the fetus (Jensen et al. 1985, 1987).

Conclusions

Sufficient fetal oxygen supply during pregnancy is the basis for normal fetal growth and development. Impaired oxygen delivery can be assumed in some clinical situations such as PIH, multiple pregnancies, and diabetes in pregnancy. Evidence for fetal borderline oxygenation is given when the fetal heart rate decreases during uterine contractions since the contraction stress test gives information on the hemodynamic buffer capacity of the uterine circulation. Early labor in complicated pregnancies gives rise to the speculation that labor at term is initiated by a fall in fetal O_2 throughout gestation. This could recently be shown by fetal blood sampling using cordocentesis (Soothill et al. 1986a, b, 1987).

References

Acheson GH, Dawes GS, Mott JC (1957) Oxygen consumption and the arterial oxygen saturation in foetal and newborn lambs. J Physiol (Lond) 135: 623

Ballew C, Haas JD (1986) Hematologic evidence of fetal hypoxia among newborn infants at high altitude in Bolivia. Am J Obstet Gynecol 155: 166–169

Bartels H, Baumann F (1972) Preplacental gas exchange. In: Longo LD, Bartels H (eds) Respiratory gas exchange and blood flow in the placenta. D. H. E.W. Publication, No NIH 73/361, pp 1–11

Block BSB, Llanos AJ, Creasy RK (1984) Responses of the growth-retarded fetus to acute hypoxemia. Am J Obstet Gynecol 148: 878–885

Carter AM (1989) Factors affecting gas transfer across the placenta and the oxygen supply to the fetus. J Dev Physiol 12: 305–322

Clapp JF (1978) The relationship between blood flow and oxygen uptake in the uterine and umbilical circulations. Am J Obstet Gynecol 132: 410–413

Clapp JF, Szeto HH, Larrow R, Hewitt J, Mann LI (1980) Umbilical blood flow response to embolization of the uterine circulation. Am J Obstet Gynecol 138: 60–67

Dawes, GS (1968) The placenta and fetal growth. In: Dawes GS (ed) Fetal and neonatal physiology. Yearbook Medical Publishers, Chicago, pp 42–59

Edelstone DI, Cinae ME, Fumia FD (1985) Relationship of fetal oxygen consumption and acid-base balance to fetal hematocrit. Am J Obstet Gynecol 151: 844–851

Emmrich P (1981) Diabetes mellitus der Mutter. In: Becker V, Schiebler TH, Kubli F (eds) Die Plazenta des Menschen. Thieme, Stuttgart, pp 359–368

Greiss FC (1965) Effect of labor on uterine blood flow. Am J Obstet Gynecol 93: 917

Gruenwald P (1969) Still birth and early neonatal death. In: Butler NR, Alberman ED (eds) Perinatal problems. Livingstone, Edinburgh, p 163

Harbert GM (1982) Effects of uterine contractions. In: Mohawad AH, Lindheimer MD (eds) Uterine and placental blood flow. pp 193–200

Hohmann M, Künzel W (1990) Hypotonie und Schwangerschaft. Gynakologe 23: 33–40

Hohmann M, Künzel W, Kirschbaum M (1986) Wehenbelastungstest mit Oxytocin-Nasenspray zur Diagnose der fetalen Hypoxämie. Z Geburtshilfe Perinatol 190: 210–214

Hughes AB, Jenkins DA, Newcombe RG, Pearson JF (1987) Symphysis-fundus height, maternal height, labor pattern, and mode of delivery. Am J Obstet Gynecol 156: 644–648

Ibsen HL (1928) Prenatal growth in guinea pigs with special reference to environmental factors affecting weight at birth. J Exp 200 l 51: 51–91

Itskovitz J, LaGamma EF, Rudolph AM (1983) The effect of reducing umbilical blood flow on fetal oxygenation. Am J Obstet Gynecol 145: 813–818

Itskovitz J, LaGamma EF, Rudolph EM (1987) Effects of cord compression on fetal blood flow distribution and O_2 delivery. Am J Physiol 252: H110–H109

Jensen A, Künzel W, Hohmann M (1985) Fetal heart rate and fetal deterioration: clinical and experimental observations. In: Künzel W (ed) Fetal heart rate monitoring. Springer, Berlin Heidelberg New York Tokyo

Jensen A, Hohmann M, Künzel W (1987) Dynamic changes in organ blood flow and oxygen consumption during acute asphyxia in fetal sheep. J Dev Physiol 9: 543–559

Junge HD, Künzel W, Klöck FK (1977) Acute reduction of uterine blood flow and fetal heart rate changes in pregnant sheep near term. J Perinat Med 5: 39

Kaufmann P (1981) Entwicklung der Plazenta. In: Becker V, Schiebler TH, Kubli F (eds) Die Plazenta des Menschen. Thieme, Stuttgart, pp 13–50

Künzel W (1973) Die Durchblutung am Ende der Schwangerschaft und während der Geburt. In: Dudenhausen JW, Saling E (eds) Perinatale Medizin, vol IV. Thieme, Stuttgart, p 127

Künzel W (1986) Fetal heart rate monitoring: clinical practice and pathophysiology. Springer, Berlin Heidelberg New York Tokyo

Künzel W (1989) Das aktuelle kindliche Risiko. In: Hillemanns HG, Schillinger H (eds) Das Restrisiko gegenwärtiger Geburtshilfe. Springer, Berlin Heidelberg New York Tokyo, pp 111–115

Künzel W, Hohmann M (1989) Grundlagen der kardiotokographischen Überwachung des Feten während der Schwangerschaft. Gynakologe 22: 156–161

Künzel W, Moll W (1972) Uterine O_2 consumption and blood flow of the pregnant uterus. Z Geburtshilfe Perinatol 178: 108

Künzel W, Junge HD, Klöck FK, Moll W (1972) The effect of low molecular weight dextran on uterine blood flow and the gas partial pressure on the fetal blood. Z Geburtshilfe Perinatol 176: 290

Künzel W, Klöck FK, Junge HD, Moll W (1974) Uterine blood flow, oxygen uptake and vascular resistance of pregnant sheep near term. J Perinat Med 2: 101

Künzel W, Mann LE, Bhakthavathsalan A, Ayromlooi J (1977) The effect of umbilical vein occlusion on fetal oxygenation – cardiovascular parameters and fetal electroencephalogram. Am J Obstet Gynecol 128: 201–208

Künzel W, Kastendieck E., Hohmann M (1983) Heart rate and blood pressure response and metabolic changes in the sheep fetus following reduction of uterine blood flow. Gynecol Obstet Invest 15: 300

Lang U, Künzel W (1989) Diabetes mellitus in pregnancy. Management and outcome of diabetic pregnancies in the state of Hesse, FRG; a five-year-survey. Eur J Obstet Gynecol Reprod Biol 33: 115–129

Longo LD, Hill EP, Power GG (1972) Factors afacting placental oxygen transfer. In: Longo LD, Bartels H (eds) Resperatory gas exchange and blood flow in the placenta. D. H. E.W. Publication, No NIH 73/361, pp 345–393

Moll W (1972) Gas exchange in concurrent, countercurrent and crosscurrent flow systems. The concept of the fetoplacental unit. In: Longo LD, Bartels H (eds) Respiratory gas exchange and blood flow in the placenta. U.S. Department of Health, Education and Welfare, Bethesda

Moll W, Künzel W (1974) Der uteroplacentare Kreislauf. Z Geburtshilfe Perinatol 178: 1

Owens JA, Falconer J, Robinson J (1987) Effect of restriction of placental growth on fetal and utero-placental metabolism. J Dev Physiol 9: 225–238

Parer JT, De Lammoy CW, Hoversland AS, Metcalfe J (1968) Effect of decreased uterine blood flow on uterine oxygen consumption in pregnant macagues. Am J Obstet Gynecol 100: 813

Paulone ME, Edelstone DI, Schedd A (1987) Effects of maternal anemia on uteroplacental and fetal oxidative metabolism in sheep. Am J Obstet Gynecol 156: 230–236
Soothill PW, Nicolaides KH, Campbell S (1987) Prenatal asphyxia, hyperlacticaemia, hypoglycaemia, and erythroblastosis in growth retarded fetuses. Br Med J 294: 1051–1053
Soothill PW, Nicolaides KH, Rodeck CH, Campbell S (1986b) Effect of gestational age on fetal and intervillous blood gas and acid-base values in human pregnancy. Fetal Ther 1: 168–175
Soothill PW, Nicolaides KH, Rodeck CH, Gamsu H (1986a) Blood gases and acid-base status of the human second-trimester fetus. Obstet Gynecol 68: 173–176
Steiner M (1989) Perinatalstudien in Deutschland. In: Hillemanns HG, Schillinger H (eds) Das Restrisiko gegenwärtiger Geburtshilfe. Springer, Berlin Heidelberg New York Tokyo, pp 101–110
Wigglesworth JS (1964) Experimental growth retardation in the foetal rat. J Pathol Bacteriol 88: 1–13
Wulf KH, Künzel W, Lehmann V (1972) Clinical aspects of placental gas exchange. In: Longo LD, Bartels H (eds) Resperatory gas exchange and blood flow in the placenta. D. H. E. W. Publication, No NIH 73/361, pp 505–525

Fetal Breathing Movements in Normoxia and in Hypoxia

G. S. DAWES

I am delighted to be asked to prepare this chapter on fetal breathing movements since it was a German obstetrician, von Ahlfeld (1888), who more than a century ago recorded what he deduced (correctly we now know) to be human fetal breathing movements from the maternal abdomen. The fact that many of his contemporaries, and others in subsequent generations, had doubts about his conclusions is irrelevant. I first learned of his work through Bill Windle (Windle 1940), with whom I later worked 1958–1965, speculated (Dawes 1968) that the last word on this subject had not yet been said, and was led to restudy the subject in 1969. This pursuit of knowledge has led my colleagues and me a fascinating dance ever since.

Hypoxia and the Arrest of Respiratory Movements

In adult animals and in man hypoxia, induced without any change of arterial carbon dioxide pressure (pCO_2) causes an increase in pulmonary ventilation through a rise in the amplitude and frequency of respiratory movements. This effect is largely due to excitation of the carotid bodies. From time to time it has been reported that under unusual circumstances a gross reduction of oxygen availability in adults may be associated with a decrease of ventilation; this has usually been attributed to a direct depression of the respiratory centres. The possibility of some other specific central action has also been mentioned (e.g. St John and Wang 1977). Nevertheless, we were surprised to find that in fetal lambs, unanaesthetized in utero, isocapnic hypoxia, induced by giving a low oxygen mixture to the mother to breathe (3% CO_2 was added to ensure isocapnia), caused an immediate reduction or arrest of fetal breathing movements (Boddy et al. 1974a).

This observation was made against a strange background. Most physiologists before 1970 believed that fetal breathing movements were unusual or abnormal before birth, but the advent of new techniques for work on unanaesthetized fetuses in utero made it possible to demonstrate that they were normally present though only episodically. In sheep, they occur only in rapid eye movement sleep associated with low-voltage electrocortical activity. They are abolished by small doses of general anaesthetics. Their presence was confirmed by many methods; by measuring the pressure changes between trachea and amniotic fluid (Dawes et al. 1972), by recording the diaphragmatic and intercostal electromyograms, by measuring the inward and outward flow of fluid through the tracheobronchial tree with an electromagnetic flow meter, by measuring the changes in transthoracic diameter with sets of implanted ultrasound transducers and receivers (Poore and Walker 1980), by observing the fetal diaphragm and chest wall with real time ultrasound (after 1976), and finally by recording impulse traffic from the phrenic nerve in utero (Hanson et al. 1987). So there is no doubt that the fetus makes breathing movements, although since the lungs are fluid filled and the face immersed in amniotic fluid, there is no transpulmonary gas exchange. In so far as pulmonary ventilation in utero affects fetal oxygenation at all, it may be reduced a fraction by the increased oxygen uptake in the muscles of respiration, and perhaps to a trifling extent by the episodic rises in umbilical venous pressure attributed to the see-saw movement of the abdomen and chest wall.

The demonstration that in the fetus mild hypoxaemia, far short of that required to produce acidaemia, induced an immediate arrest of fetal breathing movement (Boddy et al. 1974a) posed a difficult question as to mechanism. The carotid and aortic bodies of fetal lambs in utero respond to hypoxia or hypercapnia over the range of blood gas values normally encountered as far back as 90 days gestation (term in the sheep is 147 days), i.e. at an early age before behavioural states associated with sleep appear in this species (Blanco et al. 1984; Kumar and Hanson 1989). The response could not be attributed to failure of the sensory end organs which monitor arterial oxygenation postnatally. Nor was the arrest of breathing due to some anomalous response to systemic arterial chemoreceptor stimulation before birth; section of the afferent nerves, the vagi and the carotid nerves, did not abolish the arrest of fetal breathing induced by hypoxia.

A Specific Central Mechanism for Hypoxic Arrest of Breathing

In a search for the mechanism, sections were made through the upper and lower brain stem of fetal lambs; these gave interesting results (Dawes et al. 1983). In the first place, rostral transection of the brain just above the colliculi dissociated the episodic electrocortical activity from episodic breathing movements near term. Since these are normally associated, we deduced that there was a descending pathway from the centre(s) which determined electrocortical activity and which entrained the activity of the fetal respiratory centre. The location of the centre which controls electrocortical activity is still a matter of speculation.

Secondly, we found that transection at the level of the inferior colliculus or in the upper pons (i.e. more caudal than the previous lesions) caused a gross change in the response to isocapnic hypoxia. In place of an arrest or reduction of respiratory movements, there was now an increase in both amplitude and rate. The simplest explanation, but not the only possible one, was that hypoxia activated (directly or indirectly) a centre in the nervous system above the level of the lesion, which by a descending pathway caused arrest of respiratory activity. Further experiments by Parkes et al. (1984) showed that lesions placed centrally in the upper pons caused dissociation between electrocortical activity and breathing movements with no change in the response to hypoxia. After bilateral lesions placed laterally on each side of the upper pons, hypoxia no longer caused an arrest of respiratory movements; but the lesions were large, many millimeters in diameter. Meanwhile Barbara Johnston, who had taken part in the original work at Oxford, moved to New Zealand and was able to place smaller lesions in the upper pons of fetal lambs using stereotactic equipment adapted by Peter Gluckmann for work on fetal sheep. Gluckmann and Johnston (1987) described lesions close to the nucleus parabrachialis medialis which caused reversal of the normal fetal response to hypoxia; it was not clear whether the results were due to interruption of a tract or damage to a particular nucleus of cells. The localization is not yet certain; it is difficult to place accurately small bilateral lesions.

Implications for the Newborn and Adult

Further work in rats postnatally (Martin-Body and Johnston 1988; Hanson and Williams 1989) has shown that hypoxia causes a reduction in breathing movements by a pathway which can be interrupted by decerebration. This provides an explanation for the phenomenon described by Cross and Warner (1951), who showed that exposure of a newborn human infant to 15% oxygen (in place of the 21% present in normal air) caused a primary stimulation of breathing movements followed by a secondary depression. At the time Cross and his colleagues attributed this secondary depression to a generalized effect of oxygen lack on the respiratory centre. This is now attributed to specific stimulation of an area above the pons.

Martin-Body (1988) has also shown that in adult rats, after cutting the carotid nerves, hypoxia causes an immediate arrest of breathing movements which is abolished by transection of the brain stem. So, not surprisingly, in the adult as in the fetus there is a specific mechanism by which hypoxia causes arrest of breathing. Perhaps this is useful prenatally for conserving oxygen before pulmonary gaseous exchange is established. Perhaps it had an even more important purpose in our remote ancestors. We do not yet know whether such a mechanism exists in fish to modify opercular activity, or in lung fish to arrest breathing under unfavourable conditions.

Methods for Locating the Hypothetical Oxygen Sensor

Meanwhile the hunt to localize the hypothetical chemoreceptor centre continues. There are several reports of respiratory arrest from the stimulation of areas in the pons (e.g. Coles 1987), which could be due to the excitation of nonspecific descending tracts; breathing can be arrested voluntarily or during swallowing. The most promising lead comes from the work of Noble in Reading. He introduced electrodes into the rostral pons of fetal lambs and has recorded, from a set of cells located within a small distance of one another, activity which increases greatly during fetal hypoxia (Noble and Williams 1989). This is an encouraging development which opens new possibilities for the future. The normal function of these cells is as yet unknown. They could be in the pathway of a descending neural network activated by hypoxia from the pons or above, or they could be the central connections of the systemic arterial chemoreceptors. There is uncertainty as to whether their localization coincides with the areas identified by Johnston (whose lesions obliterated the response to hypoxia). I am inclined to the view that these two areas are distinct. The problem is complicated by the fact that Johnston's published sections are cut on the slant. Whatever the outcome this is now a vigorous area for research.

Hanson and this colleagues have also recently shown that almitrine, which stimulates the systemic arterial chemoreceptors in adult animals, causes arrest of breathing movements in the fetal lamb. This response is not affected by cutting the carotid nerves or the vagi, but is abolished by Barbara Johnston's lesions (Bennett et al. 1989). The simplest and possibly the most likely explanation is that almitrine excites central chemoreceptors in the lower brain stem to cause arrest of fetal breathing movements. It would appear that before birth the central mechanisms, which in the adult ensure that carotid body stimulation causes hyperventilation, are gated out. There are other analogies to consider, for example, the fact that decerebration of the fetus in utero does not cause rigidity. Whatever the outcome it seems that almitrine may prove a most useful tool.

Clinical Application

Soon after the discovery of fetal breathing, its association with rapid eye movement sleep, and its inhibition by acute hypoxaemia, we speculated that this might be a useful marker of fetal health (Boddy et al. 1974b). Indeed, Frank Manning, who spent a productive year in Oxford in the early 1970s exploring ways in which breathing movements could be recorded in the human fetus, subsequently developed a "biophysical profile" in which fetal breathing movements are one factor in assessing health. This has proved useful practically. Like the Apgar score, it has directed attention to the desirability of measuring characteristics of fetal behaviour other than the pulse rate.

However, in the last few years it has become clear that the regulation of fetal breathing is still more sophisticated. The lessons of the past 25 years have repeatedly shown that the fetus has been undervalued in the complexity of its physiological behaviour. Firstly, evidence has accumulated to show that the effect of hypoxia in decreasing fetal breathing movements is less at earlier gestational ages in sheep, whether the hypoxia was induced by reducing maternal inspired O_2 (Clewlow et al. 1983), or by reducing maternal uterine blood flow (Bocking and Harding 1986). Both of these studies, and previous observations (Boddy et al. 1974a), suggest that hypoxaemia reduces the incidence of low-voltage electrocortical activity near term, which also may partly explain the reduction of fetal breathing activity.

Secondly, evidence is accumulating of fetal adaptation to hypoxia near term. Bocking and Harding (1986) showed a partial return of breathing movements after only 2 h exposure to hypoxia in younger fetuses. Two other research groups have observed during prolonged hypoxia in older fetal lambs that after 12–24 h fetal breathing is indistinguishable from that which preceded the onset of maintained hypoxia, whether the hypoxia is caused by reducing the O_2 content of the air inspired by the mother sheep (Koos et al. 1988) or by reduction of the maternal arterial blood supply to the placenta (Bocking et al. 1988b). Explanations are still speculative. There is a rapid rise in adrenaline and haemopoietin concentrations in fetal plasma but the breathing movements return before the increase in fetal haemoglobin concentration and O_2 content (Kitanaka et al. 1989). The 2,3-DPG values are unchanged. Blood flow to the heart and brain is much increased (Bocking et al. 1988a, b), but the increase is immediate while breathing is decreased. So there may be adaptation at the cellular level, or some other central hormonal change. These findings suggest caution in clinical application, since it would seem that the presence of fetal breathing does not guarantee normoxia, though it does give evidence of normal brain stem function in compensated hypoxaemia.

References

Bennett LA, Gluckmann PD, Hanson MA, Johnston BM, Moore PJ (1989) Almitrine mimics hypoxia in fetal sheep with lateral pontine lesions. J Physiol (Lond) 417: 20P

Blanco CE, Dawes GS, Hanson MA, McCooke HB (1984) The response to hypoxia of arterial chemoreceptors in fetal sheep and newborn lambs. J Physiol (Lond) 351: 25–37

Bocking AD, Harding R (1986) Effects of reduced uterine blood flow on electrocortical activity, breathing, and skeletal muscle activity in fetal sheep. Am J Obstet Gynecol 154: 655–662

Bocking AD, Gagnon R, White SE, Homan J, Milne KM, Richardson BS (1988a) Circulatory responses to prolonged hypoxemia in fetal sheep. Am J Obstet Gynecol 159: 1418–1424

Bocking AD, Gagnon R, Milne KM, White SE (1988b). Behavioral activity during prolonged hypoxemia in fetal sheep. J Appl Physiol 65: 2420–2426

Boddy K, Dawes GS, Fisher R, Pinter S, Robinson JS (1974a) Fetal respiratory movements, electrocortical and cardiovascular responses to hypoxaemia and hypercapnia in sheep. J Physiol (Lond) 243: 599–618

Boddy K, Dawes GS, Robinson J (1974b) Intrauterine fetal breathing movements. In: Gluck L (ed) Modern perinatal medicine. Yearbook Medical Publishers, Chicago, pp 381–390

Clewlow F, Dawes GS, Johnston B, Walker DW (1983) Changes in breathing, electrocortical activity and muscle activity in unanaesthetized fetal lambs with age. J Physiol (Lond) 341: 463–476
Coles SK (1987) Mesencephalic apnoeic regions in the rat. J Physiol (Lond) 382: 176P
Cross KW, Warner P (1951) The effects of inhalation of high and low oxygen concentrations on the respiration of the newborn infant. J Physiol (Lond) 114: 283–295
Dawes GS (1968) Foetal and neonatal physiology. Yearbook Medical Publishers, Chicago
Dawes GS, Fox HE, Leduc B, Liggins GC, Richards RT (1972) Respiratory movements and paradoxical sleep in the foetal lamb. J Physiol (Lond) 220: 119–143
Dawes GS, Gardner WN, Johnston BM, Walker DW (1983) Breathing in fetal lambs: the effect of brain stem section. J Physiol (Lond) 335: 535–553
Gluckmann PD, Johnston BM (1987) Lesions in the upper lateral pons abolish the hypoxic depression of breathing in unanaesthetized fetal lambs in utero. J Physiol (Lond) 382: 373–383
Hanson MA, Williams BA (1989) The effect of decerebration and brain-stem section on the ventilatory response to acute hypoxia in normoxic and chronically hypoxic newborn rats. J Physiol (Lond) 414: 25P
Hanson MA, Moore PJ, Nijhuis JG (1987) Chronic recording from the phrenic nerve in fetal sheep in utero. J Physiol (Lond) 394: 4P
Kitanaka T, Alonso JG, Gilbert RD, Siu BL, Clemons GK, Longo LD (1989) Fetal responses to long-term hypoxemia in sheep. Am J Physiol 256: R1340–1347
Koos BJ, Kitanaka T, Matsuda K, Gilbert RD, Longo LD (1988) Fetal breathing adaptation to prolonged hypoxaemia in sheep. J Dev Physiol 10: 161–166
Kumar P, Hanson MA (1989) Re-setting of the hypoxic sensitivity of aortic chemoreceptors in the newborn lamb. J Dev Physiol 11: 199–206
Martin-Body RL (1988) Brain transections demonstrate the central origin of hypoxic ventilatory depression in carotid body-denervated rats. J Physiol (Lond) 407: 41–52
Martin-Body RL, Johnston BM (1988) Central origin of the hypoxic depression of breathing in the newborn. Respir Physiol 71: 25–32
Noble R, Williams BR (1989) Excitation of neurones in the rostral pons during hypoxia in anaesthetized neonatal lambs. J Physiol (Lond) 417: 146P
Parkes MJ, Bamford OS, Dawes GS, Gianopoulos JG, Quail AW (1984) The effects of removal of the cerebellum, brain stem transection and discrete brain lesions in fetal lambs. Proc Soc Study Fetal Physiol, Oxford 8
Poore ER, Walker DW (1980) Chest wall movements during fetal breathing in the sheep. J Physiol (Lond) 301: 307–315
St John WM, Wang SC (1977) Response of medullary respiratory neurones to hypercapnia and isocapnic hypoxia. J Appl Physiol 43: 812–821
von Ahlfeld F (1888) Über bisher noch nicht beschriebene intrauterine Bewegungen des Kindes. Verhandlungen der Deutschen Gesellschaft für Gynäcologie. Breitkopf and Härtel, Leipzig, pp 203–210
Windle WF (1940) Physiology of the fetus. Saunders, London

Maternal Respiration – Its Effect on the Fetus

A. HUCH and R. HUCH

In 1876 Paul Zweifel, then a coworker of Hoppe-Seyler in Strasbourg in the Archives for Gynecology, reported on the spectroscopic determination of the typical oxygen absorption bands in the umbilical blood before the newborn infant had taken its first breath. That was a totally new finding and completely unexpected, considering the state of knowledge at that time. Paul Zweifel gave the following interpretation: since the fetus has absolutely no contact with atmospheric air, its only source of oxygen must be the maternal blood, and oxygen transfer is thus directly proved. This transfer must occur from maternal to fetal blood through the epithel of the chorionic villi.

A century later this is all simple, common knowledge for us, and yet, paradoxically, in spite of all our sophisticated diagnostic possibilities, there is today still a degree of uncertainty resulting in contradictory viewpoints about the degree to which maternal respiration can influence fetal oxygenation. This is espicially true for respiration deviating from the norm, e.g., hyperventilation and breathing of supplemental oxygen. Experience with the transcutaneous oxygen tension (pO_2) electrode with simultaneous measurements on mother and fetus allowed a somewhat better description of the relationships (A. Huch et al. 1977, 1980).

Maternal Hyperventilation

It is a well-known fact that a significant increase in alveolar ventilation, caused by hormonal changes, already occurs in the first weeks of pregnancy and is substantially higher than the oxygen required by the mother and fetus (R. Huch and A. Huch 1984). The maternal acid base status is correctly described as a compensated respiratory alkalosis. The maternal arterial pCO_2 values are approximately 10 torr (1.3 kPa) lower than the values measured in nonpregnant females. Theoretically, assuming the validity of the alveolar equation, analogue increases for pO_2 should be found. Only a few studies, however, saw any sort of significant change in the maternal pO_2 (R. Huch and A. Huch 1984).

During labor and delivery, maternal respiration increases as a result of fear, anxiety, pain, or, sometimes, voluntarily, when respiratory instructions are followed which should direct the mother's concentration to correct breathing and away from the pain. There is general agreement that during parturition a woman breathes excessively as a result of the contractions, and that a marked relationship exists between the amount of respiration and the intensity of the contraction pain. Several studies have gone into this in detail (Bonica 1972; Cole and Nainby-Luxmoore 1962, Crawford 1972; A. Huch et al. 1974).

The tidal volume, which is already higher during pregnancy, can increase to values of over 2 l during a painful contraction; as much as 90 l respiratory minute volume have been measured. Since the dead space remains nearly the same, the alveolar ventilation increases overproportionally. Even when one takes into account the increase in oxygen consumption during delivery as a result of the physical work, an unavoidable result of the excessive alveolar ventilation, that is, hyperventilation, is a further decrease in pCO_2.

Numerous studies support the impression that pain is the main cause of the excessive hyperventilation. Tidal volume and respiratory minute volume are less after analgesia, the pCO_2 is higher and the pH less alkalotic. The consequences for maternal blood oxygenation are discussed below.

Why Should Maternal Hyperventilation Be Avoided?

Figure 1 shows the results observed for maternal hyperventilation (R. Huch 1986). It is clear that the increase in maternal pO_2 during hyperventilation, assuming a simultaneous decrease in pCO_2, positively influences the blood gas exchange by increasing the gradients between the maternal and fetal blood – as long as the perfusion of the intervillous space does not change.

Hyperventilation, however, affects the cardiovascular system. Most people are familiar with the dizziness and blackouts which occur when one voluntarily hyperventilates; these conditions are the result of reduced brain perfusion. In addition, obstetricians and midwives are familiar with symptoms of tetany, the result of a decrease in ionised calcium in the blood caused by the alkalotic pH. Hyperventilation is usually associated with a fall in blood pressure (R. Huch 1986). This must mean a net vasodilatation in contrast to known vasoconstriction in certain organs (e.g., the brain), since, according to most studies, cardiac output increases during hyperventilation. This fall in blood pressure, along with the vasoconstriction of the uterine vessels which has been proved in animal studies, results in a decrease in uterine perfusion. The unavoidable consequences for the fetus are also listed in Fig. 1. The increase in maternal O_2 affinity, which occurs with alkalosis, has the same effect: the oxygen transfer to the fetus is more restricted. Studies in humans have shown that whenever these negative influences on the fetus are measurable, the maternal pCO_2 values must be lower than 20 mmHg (R. Huch 1986).

Findings from studies where mother and fetus were simultaneously monitored during parturition by transcutaneous pO_2 have shown another area of maternal-fetal risk associated with hyperventilation, namely the risk

↑ Δ Po_2 Mother - Fetus

↑ Δ Pco_2 Mother - Fetus

↑ Respiratory work

↓ Cerebral perfusion

↑ Tetany, Carpopedal spasms

	Fetus
↓ RR	
↓ Uterine perfusion	↓ pH
↓ O_2 affinity	↓ Po_2
↓ Po_2 between contractions	↑ Pco_2

Fig. 1. Consequences of maternal hyperventilation

of maternal hypoventilation during the phases between contractions (A. Huch et al. 1974, 1977).

As is known, respiration is regulated by chemoreceptors; deviations of blood pO_2 and pCO_2 from the norm trigger a response with stronger or weaker "firing" signals to the respiratory center. After artificial hyperventilation, an apneic response is common (Kellog 1964). The respiratory activity stops totally or partially until the pCO_2 is again in the normal range. Pearl divers in Japan make use of this fact: before diving, they hyperventilate sufficiently so that the CO_2 pressure remains low, and the stimulus to breathe does not occur for a relatively long time.

How regular respiration is in the pause between two contractions is, therefore, dependent on the extent of hyperventilation during a painful contraction. In Fig. 2 the ventilatory increase is moderate. In this thoracic impedance recording, the respiratory increase during the contraction and the reduction in the pause can both be seen, yet no definite apnea phases occur. The maternal pO_2 curve looks like another contraction curve. There is no indication that these slight pO_2 fluctuations represent any disadvantage to the fetus. At this pressure, the maternal blood is nearly fully saturated with oxygen. On the other hand, apnea phases after excessive hyperventilation have been observed (Fig. 3). They are often as long as the pause between contractions. When the fetus is simultaneously monitored for pO_2 (the electrode being placed on its presenting part), it can be seen that these maternal pO_2 decreases are large enough to affect fetal oxygenation (A. Huch et al. 1974).

Pethidine has a depressive effect on respiration; the extent of hypoventilation in the pause between contractions and thus the extent of the pO_2 decrease are larger after administration of pethidine (A. Huch et al. 1974).

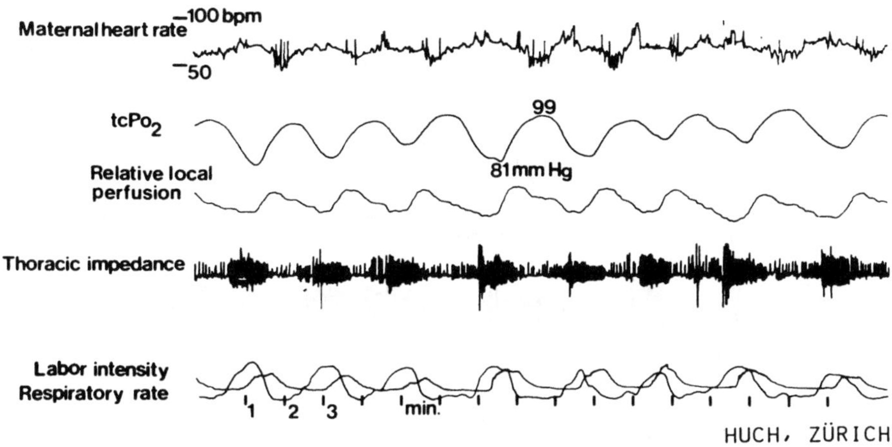

Fig. 2. Excerpt of continuous recordings of maternal heart rate, $tcpO_2$, transthoracic impedance and contractions during labor. Moderate increases in ventilation during contractions result in a fluctuating pO_2 curve (from A. Huch et al. 1980)

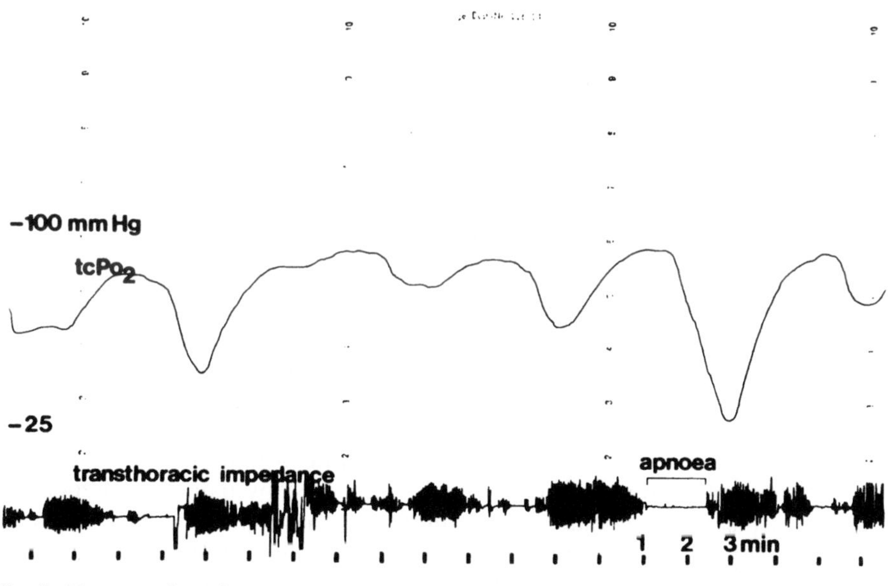

Fig. 3. Excerpt of continuous recordings of maternal $tcpO_2$ and contractions during labor. Apnea phases in the pause between contractions are reflected in significant $tcpO_2$ decreases (from A. Huch et al. 1974)

The lowest and mean PO_2 values in each maternal transcutaneous pO_2 recording after the administration of 50 mg or 100 mg pethidine were lower than before administration (A. Huch et al. 1974). It is thus possible that not only the pethidine transfer to the fetus, but also its direct effect on the maternal respiratory center are responsible for the newborn respiratory depression which is sometimes observed.

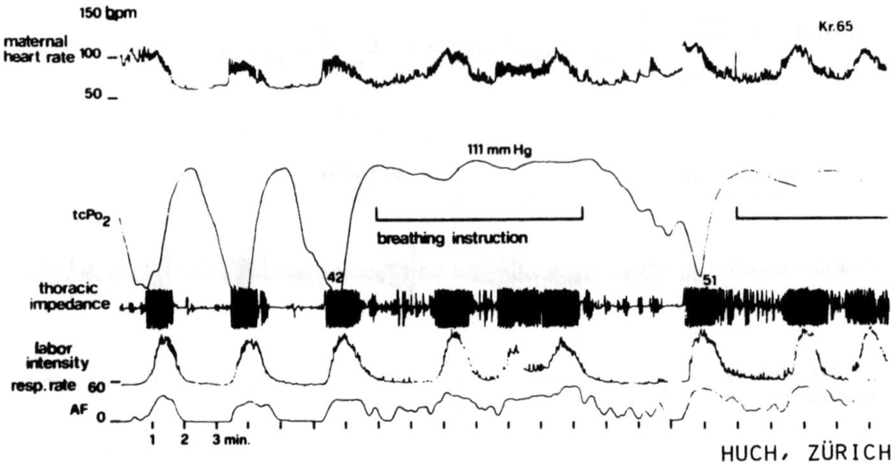

Fig. 4. Same as in Fig. 1. Breathing instructions during the pause between contractions result in high and stable maternal $tcpO_2$ values (from R. Huch and A. Huch 1984)

Fortunately, something can be done to counter these maternal pO_2 decreases, namely instructive respiratory guidance during the pauses between contractions. Even if it is not always possible to prevent hyperventilation during contractions, respiratory instructions and guidance in the pauses can keep the pO_2 from decreasing (Fig. 4).

That it is mainly pain which causes hyperventilation during contractions has already been well shown by several studies and with several respiratory parameters. Strasser et al. (1975) also showed the same by studying transcutaneous pO_2 measurements. Before complete analgesia was reached under peridural anesthesia, a maximal pO_2 decrease of 56 mmHg was seen; after analgesia, when there was no sensation of pain, stable and equally high values of pO_2 were noted during contractions as were seen in the pauses between the contractions. In summary, it can be said that transcutaneous pO_2 measurements have confirmed that the pregnant woman during parturition has a tendency to hyperventilate which is dictated by pain. This hyperventilation is disadvantageous for the fetal oxygen supply and must thus be avoided. Any method or technique which reduces or totally eliminates pain can therefore only be of benefit to mother and child as long as there are no negative side effects.

The use of the transcutaneous pO_2 technique during parturition on both mother and fetus has also brought somewhat more objectivity into the controversy over the use of supplemental oxygen. There is still no consensus on the administration of oxygen during labor and delivery. The main arguments raised against supplemental oxygen administration have been:

1. Additional oxygen causes a contraction of the uterine vessels and therefore may lower fetal PO_2 and pH, particularly if administered over a long period.

2. Fetal pO_2 increases only when it is initially high.
3. The reported small increases in fetal pO_2 are not considered to be significant to fetal oxygenation.
4. There is an inherent risk of oxygen toxicity to the fetus.

Our own results have been compiled from a total of 123 periods of oxygen administration to 69 women in labor (R. Huch et al. 1979). The mean length of the oxygen or oxygen-air breathing period was 14 min, with a range of 3–80 min. Transcutaneous pO_2 was recorded both from the mothers and their fetuses. Maternal and fetal heart rates as well as contractions were also recorded.

Figure 5 summarizes the results from all 123 periods of maternal oxygen administration. The mean increase in fetal transcutaneous pO_2 was 5 mmHg (from 16.5 to 21.5 mmHg), whereas maternal transcutaneous pO_2 was 88.5 mmHg before and about 300 mmHg during supplementary oxygen breathing. It was clearly documented that, despite significant variations in the initial PO_2

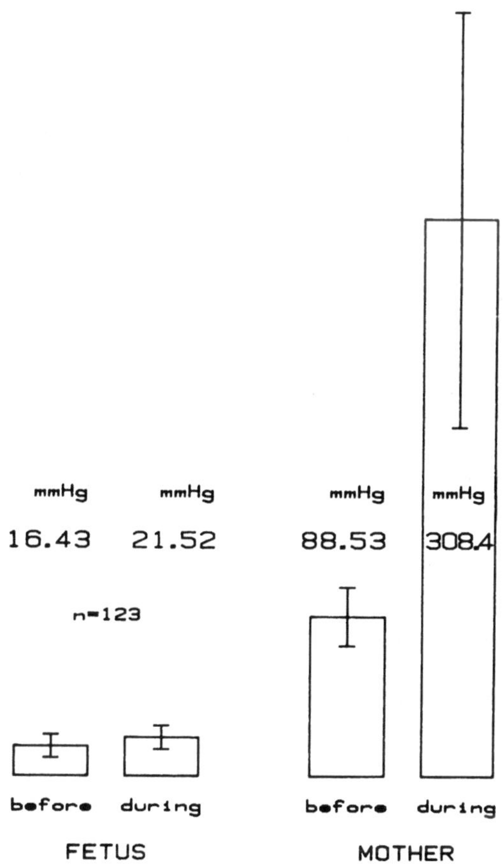

Fig. 5. Mean fetal and maternal $tcpO_2$ (+ SD) values before and during supplemental maternal oxygen

levels, fetal pO_2 rose in nearly each case. It was particularly noteworthy that there was an increase even with very low initial transcutaneous pO_2 and that in certain individual cases there were large increases.

Analysis of 12 recordings where maternal supplemental oxygen breathing ranged from 20 to 80 min showed that fetal transcutaneous PO_2 increased independently of the length of oxygen administration to the mother. There was no tendency of transcutaneous pO_2 to fall during the long periods of oxygen breathing. This is in agreement with results from studies with animals. Battaglia (1979) studied the uterine blood flow in sheep during supplemental oxygen administration to the mother and found no decrease in blood flow. Matalon et al. (1978) gave supplementary oxygen to pregnant sheep, and saw an increase in fetal pO_2 in all cases; in no instance did they note any fall in either maternal or fetal pH.

A mean increase in fetal pO_2 of 5 mmHg during maternal O_2 breathing may be considered to be small, particularly when compared with the large increase in maternal pO_2. This has frequently been cited as proof that the uterine vessels contract. This misinterpretation is due to the fact that only pO_2 is considered. If, instead, one looks at oxygen saturation or the total oxygen content of the blood, a totally different, and much more accurate, picture emerges. With an increase in fetal pO_2 from 16.5 to 21.5 mmHg, fetal saturation increases from 30% to about 50%. Expressed in milliliters of O_2, this would be an increase of 4 ml O_2, producing an increase in oxygen availability to the fetus of as much as 58%. With an initial fetal pO_2 of, for example, only 10 mmHg and again a transcutaneous O_2 increase of 5 mmHg resulting from maternal oxygen administration, the percentage increase in oxygen availability is even greater than with a normal fetal pO_2. The benefit of maternal supplemental oxygen to the fetus was documented by our material also in the behavior of the fetal heart rate, particularly in phases with low initial fetal pO_2 levels. Seven situations with late decelerations were observed, which in two cases could be totally corrected and in three cases partly corrected by O_2 administration to the mother. There is thus no clinical or experimental evidence suggesting any risk to the fetus from maternal supplementary oxygen administration. If supplementary oxygen is withheld, however, the risk of fetal hypoxia and the resulting lesions cannot be ignored.

References

Battaglia FC (1979) Perinatal management of the very low birth weight infant. VI. Corso internazionale teorico-pratico di terapia intensiva neonatale per il binomio neonatalogo-caposala

Bonica JJ (1972) Obstetric analgesia and anesthesia. Springer, Berlin Heidelberg New York

Cole PV, Nainby-Luxmoore RC (1962) Respiratory volumes in labour. Br Med J 1: 1118–1127

Crawford JS (1972) Principles and practice of obstetric anaesthesia. Blackwell, Oxford
Huch A, Huch R, Lindmark G, Rooth G (1974) Maternal hypoxaemia after pethidine. J Obstet Gynaecol Br Cwlth 81: 608–614
Huch A, Huch R, Schneider H, Peabody J (1980) Experience with transcutaneous PO_2 ($tcPO_2$) monitoring of mother, fetus and newborn. J Perinat Med 8(51): 51–72
Huch A, Huch R, Schneider H, Rooth G (1977) Continuous transcutaneous monitoring of fetal oxygen during labour. Br J Obstet Gynaecol 84(1): 1–37
Huch R (1986) Maternal hyperventilation and the fetus. J Perinat Med 14(3): 3–17
Huch R, Huch A (1984) Maternal and fetal acid-base balance and blood gas measurement. Fetal physiology and medicine: the basis of perinatology. Marcel Dekker, New York
Huch R, Huch A, Rooth G (1979) Use of transcutaneous oxygen measurement in the perinatal period – measurement in the fetus. IXth world congress of gynecology and obstetrics. Elsevier, Amsterdam (International Congress Serie vol 512)
Kellogg R (1964) Central chemical regulation of respiration. Handbook of physiology, Sect 3: Respiration. Waverly Press, Baltimore
Matalon SV, Manning PJ, Bernie BJ, Eichhorst BC, Hunt CE, Seeds AE (1978) The effects of changes of maternal paO_2 and $paCO_2$ on the fetal paO_2 and $paCO_2$ – in vivo Study. Respir Physiol 32: 51–59
Strasser K, Huch A, Huch R (1975) Der Einfluß der lumbalen Epiduralanästhesie unter der Geburt auf die Atmung und den kontinuierlich transcutan gemessenen pO_2 der Mutter. Perinatale Medizin VI. Thieme, Stuttgart
Zweifel P (1876) Die Respiration des Fötus. Arch Gynaekol 9: 291–305

Abstracts

Influence of Maternal Oxygen Inhalation on Uteroplacental and Fetal Perfusion Measured by Doppler Sonography in Patients with Preterm Labor

K.-E. Ruckhäberle, R. Robel, R. Faber, B. Viehweg, and M. Winkler

Introduction

It was the aim of the present study to confirm and to pursue our previous nuclear medical investigations (Ruckhäberle et al. 1986) by using the Doppler ultrasound method to measure the uteroplacental, fetoplacental and fetal blood flow velocity waveforms in preterm labor. Apart from confirming impaired hemodynamics, we were especially interested to establish whether maternal oxygen inhalation could bring about any improvement of the impairment and result in a better pregnancy outcome.

Patients and Methods

In a randomized prospective study, we administered betamimetics i.v. to 54 patients with preterm labor (average gestational age 30 weeks) to achieve tocolysis. While in 21 patients tocolysis was induced in this way alone, 6 patients also received oxygen (8–10 l oxygen/min in a half-open system, corresponding to a respiratory oxygen fraction of 0.7–0.8) for 30 min, and perfusion was measured before and immediately after inhalation. In a group of 27 patients, the O_2 administration was continued for 2 weeks (30 min daily). Using a Toshiba 38 D duplex Doppler device we measured resistance index (RI) and pulsatility index (PI) in both the uterine arteries, the fetal umbilical artery, and the thoracic aorta as well as in the internal carotid artery on admission, after the first and second week. We used 30 women with a normal course of pregnancy and comparable gestational age as a control group.

Results

Considering the median values there was a tendency of increased RI and PI only in the maternal and fetal vessels (especially the PI of the uterine artery on the placental side and the RI of the fetal thoracic aorta) in preterm labor caused by having only about one-third of the real prepathological and pathological values. Under the direct influence of maternal oxygen inhalation, we demonstrated in six patients the improvement in maternal and fetal perfusion by decreased Doppler indices. In contrast to the findings in patients in whom tocolysis alone was induced patients with preterm labor who had been additionally treated with O_2 daily showed a significant decrease after 1 and 2 weeks only where there were disturbed values of PI in the uterine artery on the plancental side, PI in the umbilical artery and RI and PI in the thoracic aorta as well as a significant increase of RI in the fetal internal carotid artery. The increased PI and RI during O_2 treatment in the subgroup with normal primary values did not exceed the range of normal controls. A disturbed uterine perfusion as well as a lower RI and PI in the fetal thoracic aorta on admission are strongly related to a shorter prolongation of gestation, lower gestational age at delivery, and a lower birth weight. Additional oxygen therapy has a significant positive influence on all clinical parameters (prolongation of gestation 21.5 : 50.5 days; gestational age 33.6 : 37.2 weeks; birth weight 2295 : 2890 g) in patients with normal as well as disturbed maternal perfusion. We were encouraged to continue with the clinical administration of maternal oxygen inhalation from these results.

Reference

Ruckhäberle K-E, Vogtmann C, Petzold J et al. (1986) Fetal cardiographic and mechanocardiographic findings and uteroplacental perfusion in preterm labor. In: Grauel L, Syllm-Rapoport I, Wauer R (eds) Research in Perinatal Medicine. Thieme, Leipzig, p 307

Interpretation of FHR Patterns of SGA Fetuses in Correlation to Fetal Blood Gas Values Obtained Via Cordocentesis

E. Cetin, K. H. Nicolaides, and B. Arabin

In the management ofd pregnancies with small for gestational age (SGA) fetuses, the aim is to distinguish between normal small babies, not at risk of perinatal death or chronic handicap, and fetuses whose growth is retarded due to either placental insufficiency with malnutrition and hypoxia, or intrinsic fetal pathology like chromosomal abnormalities, or hereditary malformations.

Placental failure to support fetal growth results in reduced placental transfer of oxygen and essential nutrients, leading to reduced fetal pO_2, glucose, and essential amino acids. Subsequently there is failure to excrete CO_2 and lactate, resulting in progressive respiratory and metabolic acidosis with further disturbances in carbohydrate, lipid, and protein metabolism. For the fetus affected by intrauterine growth retardation (IUGR), antenatal surveillance is mainly aimed at detecting "fetal distress" and determining the optimum time, mode, and place of delivery®. Although fetal heart rate (FHR) monitoring is the most widely used noninvasive method of fetal oxygenation, there is considerable controversy regarding the interpretation of FHR patterns. This is evidenced by the publication of at least 25 classifications of FHR analysis. The aim of this study is to correlate various antepartum FHR parameters (performed immediately before cordocentesis) with fetal umbilical venous blood gases in SGA fetuses.

Method

In the total group of 65 SGA pregnancies, ultrasound examinations excluded fetal malformation. Cases with an abnormal karyotype were not included. The abdominal circumference (AC) was below the 5th and the head circumference (HC)/AC ratio above the 95th centile for gestational age. In 73% the amniotic fluid volume was either decreased or oligohydramnios. FHR patterns were examined using a minute-to-minute visual analysis for baseline heart rate, oscillation amplitude, and frequency accelerations (> 15 bpm/15 s) and decelerations (> 10 bpm/60 s).

Results

A low baseline variability and a reduced number of accelerations were significantly correlated with decreased pH, pO_2, oxygen content, and oxygen saturation levels (Fig. 1). No correlations were found to pCO_2 levels (Table 1). This might be explained by the fact that only umbilical vein samples

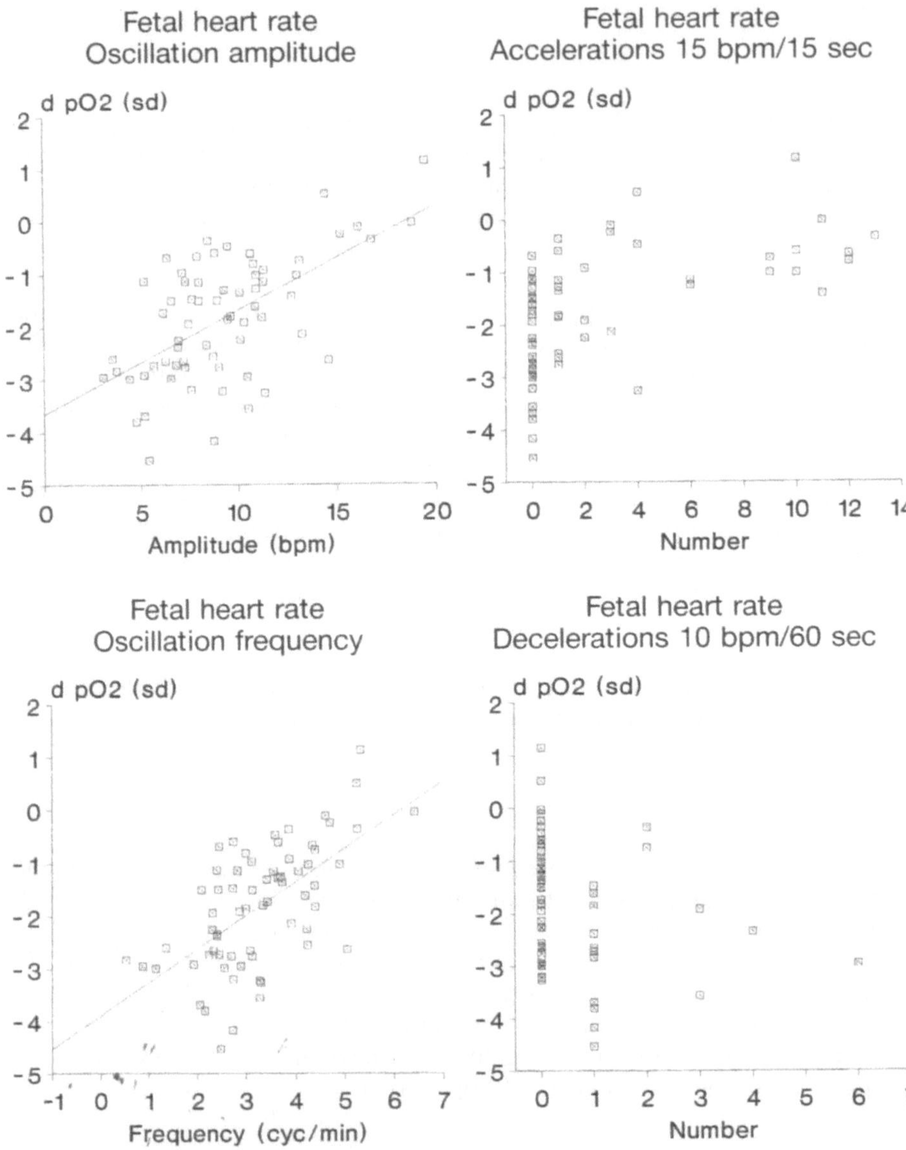

Fig. 1. FHR parameters in correlation to the degree of hypoxia (in comparison to the normal range for gestational age in standard deviations, *sd*) in umbilical venous blood

Table 1. Correlation of fetal heart rate patterns to fetal blood gas values

		pO_2	pH	pCO_2	Hb	O_2 saturation	O_2 content
Baseline heart rate	$r =$	− 0.272	− 0.059	− 0.031	0.015	− 0.340	− 0.386
	$p =$	0.073	0.640	0.530	0.212	0.005	0.014
Oscillation amplitude	$r =$	0.470	0.584	− 0.021	0.893	0.338	0.309
	$p =$	0.001	0.000	0.958	0.017	0.006	0.012
Oscillation frequency	$r =$	0.427	0.581	− 0.243	0.191	0.310	0.395
	$p =$	0.003	0.006	0.087	0.123	0.012	0.001
Acceleration	$r =$	0.364	0.410	− 0.186	0.120	0.333	0.387
	$p =$	0.003	0.000	0.122	0.336	0.006	0.001
Deceleration	$r =$	− 0.173	− 0.141	− 0.138	− 0.172	− 0.178	− 0.127
	$p =$	0.166	0.261	0.242	0.166	0.153	0.029

were included, indicating placental clearance of hypercapnia. Decreased baseline variability and a decreased number of accelerations were more strongly associated with severe degrees of hypoxia or acidemia (Fig. 1; Table 1). Furthermore, no correlations to blood gases were found with the number of decelerations, and the only significant correlation to baseline heart rate was with oxygen saturation and content.

In conclusion the placental support of the fetus, e.g., with oxygen, influences the FHR patterns only in severe cases of placental insufficiency. Therefore, we suppose that FHR monitoring is of restricted clinical value as an early scoring test for fetal malnutrition.

Reference

Nicolaides KH, Economides DL, Thorpe-Beeston (1989) Treatment of fetal growth retardation. In: Sharp F, Fraser RB, Milner RDG (eds) Fetal growth proceedings of the Twentieth Study Group of the Royal College of Obstetriciens and Gynecologists. RCOG, London

II Acute Oxygen Deficiency During Pregnancy and Birth. Influence on the Fetus

Oxygen Supply to the Fetus and Fetal Hypoxemia

A. M. Rudolph

Oxygen Delivery

The fetus receives its oxygen supply across the placenta, to which oxygen is delivered by maternal blood. The amount of oxygen delivered to the placental site is determined by the magnitude of blood flow and the oxygen content of uterine arterial blood. Transfer of oxygen from maternal to fetal blood is influenced by the diffusion gradient for oxygen between the maternal uterine blood and fetal umbilical blood. The diffusion gradient is affected by the thickness of the placental tissues separating the uterine and umbilical bloods and the relationship of arteriovenous flow patterns in the uterine and umbilical circulations. Uptake of oxygen by the fetus at the placental site is related to umbilical–placental blood flow and oxygen content of umbilical venous blood, or

$$\text{fetal } O_2 \text{ delivery} = \dot{Q}_{UV} \times \dot{C}_{UVO_2}$$

where $\dot{Q}_{UV}$ is umbilical–placental blood flow and $\dot{C}_{UVO_2}$ is umbilical venous oxygen content.

Studies in sheep have shown that the oxygen tension (pO_2) of arterial blood in the mother is about 100 mmHg, and pO_2 of uterine venous blood is about 38–40 mmHg. In the fetus, umbilical arterial pO_2 is about 20–22 mmHg, and umbilical venous pO_2 about 30–35 mmHg. Oxygen uptake by the fetus is, however, facilitated by differences in the oxygen equilibration curves for fetal and adult bloods. The oxygen equilibration curve relates the oxygen tension (on the abscissa) to the saturation of hemoglobin with oxygen (on the ordinate). The curve of blood that has a greater affinity for oxygen will be located to the left of the curve for blood with lower affinity for oxygen. The position of this curve is expressed by the P_{50}, which expresses the pO_2 at which hemoglobin is 50% saturated with oxygen. In the sheep, the P_{50} for maternal blood is 35–40 mmHg, whereas that for fetal blood is 25–28 mmHg. Thus, the oxygen saturation in maternal arterial blood at pO_2 100 mmHg is 100%, and in uterine venous blood at pO_2 38–40 mmHg is about 45%–50%, whereas in umbilical venous blood at pO_2 30–35 mmHg, it is 85%–90%, and in umbilical arterial blood at pO_2 20–22 mmHg, about 50%.

Another important factor influencing the uptake of oxygen is the hemoglobin concentration, because this determines the actual oxygen content of blood. Oxygen content is defined as the amount of oxygen attached to hemoglobin plus the dissolved oxygen. At a pO_2 of 100 mmHg, dissolved oxygen is only about 0.3 ml/dl, relatively insignificant compared with the much larger amount attached to hemoglobin. The maximum amount of oxygen that can be attached to hemoglobin, termed the oxygen capacity, is about 1.35 ml/g hemoglobin, and this is reached at pO_2 of 100 mmHg for both adult and fetal blood. Thus, differences in hemoglobin concentration could account for differences in oxygen content of fetal and maternal blood, in addition to differences in P_{50}. In sheep, there is little difference between maternal and fetal blood hemoglobin concentration, but in some species, such as the human, fetal blood hemoglobin concentration is considerably higher than that in the mother.

Thus in sheep, with a fetal blood hemoglobin concentration of 9–10 g/dl and an oxygen saturation of 85%–90%, the oxygen content of umbilical blood is 1.35 × (7.65–9.00) ml/dl plus about 0.25–0.30 ml/dl dissolved oxygen. The oxygen content of umbilical arterial blood, with a pO_2 of about 20 mmHg and oxygen saturation of 50%, is 1.35 × (4.5–5.0) ml/dl plus about 0.06 ml/dl dissolved oxygen (dissolved oxygen increases linearly with pO_2). Oxygen delivery to the fetus, as expressed by the equation above, is determined by the oxygen content of umbilical venous blood, an average of about 11.5 ml/dl, and umbilical blood flow, which in the sheep is about 200 ml/kg fetal body weight per minute. Thus, oxygen delivery is about 23 ml/kg fetal weight per minute. Not all oxygen delivered by blood to the tissues is removed; the proportion of that taken up to that delivered is termed the oxygen extraction. In the sheep fetus, as in the adult, the oxygen extraction for the whole body is about one-third (33%) of the oxygen delivered.

Oxygen consumption refers to the actual amount of oxygen taken up by the body. In the fetus, it is represented by the difference between oxygen delivered by the umbilical venous blood and that returning to the placental site through the umbilical arteries. Because umbilical arterial and venous flows are equal, oxygen consumption ($\dot{V}O_2$) is expressed by the equation

$$\dot{V}_{O_2} = \dot{Q}_{UV}\,(\dot{C}_{uvO_2} - \dot{C}_{uaO_2})$$

where $\dot{Q}_{UV}$ is umbilical–placental blood flow and $\dot{C}_{uvO_2}$ and $\dot{C}_{uaO_2}$ are the oxygen contents of umbilical venous and umbilical arterial bloods. In the sheep fetus, oxygen consumption is about 8 ml/kg body weight per minute.

Mechanisms of Reduced Oxygen Delivery

Oxygen supply to the fetus may be reduced by decreasing the amount of oxygen delivered to the placental site from the maternal circulation, by limitation of placental transfer, or by interference with uptake into or transport by the umbilical–placental circulation. Reduced oxygen delivery to the placenta could result from a fall in maternal arterial oxygen content due to hypoxemia or anemia, or from a decrease in uterine blood flow. Oxygen transfer across the placenta may be restricted by decreased diffusion resulting from placental abnormalities, or by small placental size. Reduced fetal uptake of oxygen could result from inadequate fetal umbilical–placental blood flow due to umbilical cord compression, from a decrease in fetal hemoglobin concentration, or from an abnormality of hemoglobin, which reduces its affinity for oxygen. The responses of the fetus to acute hypoxia in the human have largely been assessed by examining changes in heart rate and heart rate variability. Detailed information regarding the cardiovascular responses of the fetus to acute lack of oxygen supply has been derived from studies in fetal lambs. In early studies, the effects of reducing arterial oxygen content by administering low oxygen gas mixtures to the ewe were examined (Cohn et al. 1974). The responses of fetal lambs during late gestation (115–140 days, term = 145 days) were characterized by bradycardia, mild arterial hypertension, and a modest decrease in combined ventricular output. Distribution of the cardiac output and the blood flows to individual organs were markedly altered. Thus, umbilical–placental blood flow did not change, but blood flow to the fetal body fell by about 40%. Blood flows to the heart, brain, and adrenal gland increased considerably, but flows to other organs fell, the degree depending on severity of fetal hypoxemia. With mild hypoxemia, flow was reduced modestly to the peripheral circulation and to the lungs; during more severe hypoxemia, renal and gastrointestinal blood flow was also reduced.

Following these reports, it has generally been assumed that the responses of the fetus to oxygen supply reduction by any means were similar. To determine whether this is so, we have examined the effects of various types of reduced oxygen delivery.

In this presentation, the effects of acute reduction of oxygen delivery to the fetus by three of the different mechanisms described above have been studied in fetal sheep. It was elected to compare the effects of reducing oxygen delivery to the fetus by 50% by:

a) reducing maternal oxygen content by administering a low oxygen gas mixture to the ewe as described above (Cohn et al. 1974);
b) reducing uterine blood flow by means of a snare placed around the middle sacral artery in the ewe (the uterine arteries originate from this vessel) (Jensen et al. 1991); and
c) compressing the umbilical cord by means of an inflatable balloon applied around it (Itskovitz et al. 1987).

The oxygen delivery to the fetus is the product of umbilical blood flow and umbilical venous oxygen content. It was thus readily possible to achieve the 50% reduction in oxygen delivery by decreasing either umbilical blood flow or umbilical venous oxygen content by 50%. With both maternal hypoxemia (MH) and reduced uterine blood flow, umbilical–placental blood flow does not change significantly; therefore, by repeatedly measuring umbilical venous oxygen content, adjustments could be made to reduce oxygen delivery by 50%. To monitor umbilical–placental blood flow, an electromagnetic flow transducer was placed on the common umbilical artery. A balloon occluder around the umbilical cord was progressively inflated while monitoring umbilical blood flow continuously to reduce it to 50% of control levels. All these studies were performed in chronically instrumented fetal lambs. Combined ventricular output and organ blood flow measurements were achieved by the radionuclide-labeled microsphere technique and were made during the normoxia period and within 30 min after reducing oxygen delivery.

MH resulted in a decrease in fetal arterial pO_2 but no significant change in pCO_2, and pH fell slowly. Reducing uterine blood flow resulted in a similar fall of fetal arterial pO_2, but a rise in pCO_2 and a slow fall of pH. Umbilical blood flow reduction was associated with a lesser fall in arterial pO_2 and no significant changes in pCO_2 or pH.

General Cardiovascular Effects

Heart rate fell by about 15% with each of the three mechanisms for reducing oxygen delivery. Arterial mean pressure increased modestly with MH and reduced uterine flow, but only slightly with reduced umbilical flow. Combined ventricular output decreased to a similar extent with all three mechanisms. However, with MH and reduced uterine blood flow, there was no change in umbilical blood flow, but a fall in blood flow to the fetal body, whereas with umbilical blood flow reduction, the reduction in combined ventricular output was related to the fall in umbilical–placental flow, but total fetal body flow did not change.

Hepatic and Ductus Venosus Blood Flows

In the normoxic fetus, about 45% of umbilical venous blood passes through the hepatic circulation, whereas about 55% passes through the ductus venosus (Edelstone et al. 1978). During MH and reduced uterine blood flow, the umbilical venous return did not change; however, a greater proportion (60%–65%) passed through the ductus venosus, but the liver received only 35%–40%. Total blood flow to the liver was thus reduced considerably, and because the oxygen content of umbilical venous blood was reduced, oxygen delivery to the liver fell markedly. With reduced umbilical blood flow, there was an even greater change in distribution of umbilical venous blood, with 70%–80% passing through the ductus venosus and only 20%–30% being directed through the liver. The distribution of this small percentage of the reduced umbilical blood flow to the liver resulted in a dramatic fall in the volume of umbilical venous blood passing through this organ, particularly to the right lobe; oxygen delivery was also greatly reduced.

Organ Blood Flows

Changes in blood flows to body organs were similar during the decreased fetal oxygen delivery induced by MH and by reduced uterine blood flow. Small differences in some organs were observed; thus, peripheral blood flow fell less during reduced uterine blood flow than during hypoxemia. This could possibly be accounted for by the fact that fetal arterial pCO_2 was higher during reduced uterine blood flow, and this may have exerted a mild vasodilator effect to counter the hypoxic vasoconstriction.

Blood flow to the peripheral circulation during MH fell dramatically to about one-third of normoxemic values, whereas with reduced umbilical blood flow, it did not change significantly (Itskovitz et al. 1987). Similar differences were noted in lower body organs; blood flows to the kidneys, gastrointestinal tract, and spleen all fell with MH and reduced uterine blood flow, but did not change significantly during reduced umbilical flow. Flow to the myocardium and to the brain increased markedly during MH and reduced uterine blood flow, but rose to a lesser degree with reduced umbilical blood flow. Blood flow to the lung fell similarly in all three types of reduced oxygen delivery.

Mechanisms Responsible for Differences in Response

The increases in myocardial and cerebral blood flows during hypoxemia are thought to be related to local vascular effects of the reduced blood oxygen content. The decrease in pulmonary blood flow is also induced by local effects of the lowered oxygen content of blood perfusing the fetal lung.

However, the decrease in blood flow to the peripheral circulation, gastrointestinal tract, and kidneys is the result of chemoreceptor reflex stimulation, because after sinoaortic reflex deafferentiation there is a greatly diminished vasoconstrictor response to hypoxia in the peripheral circulation (Itskovitz et al. 1991). Circulating catecholamine concentrations are also increased and probably contribute to the peripheral vasoconstriction.

Both the more marked vasodilation of the cerebral and myocardial circulations and the greater vasoconstriction of the peripheral circulation can be at least partly explained by the differences in the changes in ascending aortic pO_2 or oxygen saturation and the degree of stimulation of the aortic and carotid chemoreceptors. During both reduced uterine blood flow and MH, umbilical venous oxygen saturation is reduced by about 50%. The umbilical venous blood directed through the ductus venosus to the inferior vena cava, and then preferentially through the foramen ovale to the left atrium, left ventricle, and aorta, thus produced a marked fall in oxygenation of ascending aortic and carotid arterial blood. During uterine blood flow reduction, carotid arterial pO_2 fell from 24.1 to 16.3 and oxygen saturation fell from 67% to 29%, a saturation change of 38%. During reduced umbilical blood flow umbilical venous pO_2 and oxygen saturation did not fall significantly, and with the maintenance of ductus venosus versus liver blood flow, the fall in carotid arterial oxygenation was considerably smaller. Thus, during 50% reduction in umbilical blood flow, carotid arterial pO_2 fell from 23.1 to 17.1 and oxygen saturation fell from 62% to 40%, a drop of only 22%.

These differences in organ blood flow changes with different mechanisms of reduced oxygen delivery are of great interest, because they will result in differences in tissue oxygen, as well as energy substrate, supplies. It is to be stressed, however, that this report deals with acute responses. Whether these differences are similar, and sustained during more chronic hypoxemia, has yet to be examined.

References

Cohn HE, Sacks EJ, Heymann MA, Rudolph AM (1974) Cardiovascular responses to hypoxemia and acidemia in fetal lambs. Am J Obstet Gynecol 120: 817–824

Edelstone DI, Rudolph AM, Heymann MA (1978) Liver and ductus venosus blood flows in fetal lambs in utero. Circ Res 42: 426–433

Itskovitz J, LaGamma EF, Rudolph AM (1987) Effects of cord compression on fetal blood flow distribution and O_2 delivery. Am J Physiol 252: H100–H109

Itskovitz (Eldor) J, LaGamma EF, Bristow J, Rudolph AM (1991) Cardiovascular responses to hypoxemia in sinoaortic-denervated fetal sheep. Pediatr Res 30: 381–385

Jensen A, Roman C, Rudolph AM (1991) Effects of reducing uterine blood flow on fetal blood flow distribution and oxygen delivery. J Dev Physiol 15: 309–323

The Role of the Sympathetic Nervous System in Preventing Brain Damage During Perinatal Asphyxia

A. JENSEN

Introduction

Perinatal asphyxia is one of the major causes of cerebral hemorrhage, which is the most common form of brain damage in the neonatal period (Pape and Wigglesworth 1979). Since a number of clinical studies have shown that babies born prematurely are much more susceptible to cerebral hemorrhage than mature babies born near term, it is likely that there are defense mechanisms against the adverse effects of asphyxia on the integrity of the fetal brain that mature during development.

These protective maturational processes are related in part to structural changes in the developing brain. There is, for example, an anatomical structure unique to the immature brain called the germinal matrix. This structure, which can only be detected histologically until 32 weeks of gestation, is a highly vulnerable subependymal tissue containing a dense network of glioblasts and fragile capillaries that cover caput and corpus of the caudate nucleus (Pape and Wigglesworth 1979). In very immature babies, this germinal matrix is extremely sensitive to asphyxia in general, and to the associated changes in both arterial blood pressure and blood flow in particular. Due to the gradual dissipation of the germinal matrix during development, the fetal brain becomes increasingly resistant to asphyxial damage (Jensen and Schumacher 1987). Other maturational processes which protect the integrity of the developing brain include cardiovascular, neurohormonal, and metabolic mechanisms. One of these is related to the maturation of the fetal sympathetic nervous system, which provides protection for the fetal brain in various ways.

The following review first describes the clinical situation, i.e., the incidence of cerebral hemorrhage, and the risk factors involved in a large population of human neonates (n = 5301), screened for cerebral abnormalities using cranial ultrasound. The second part deals with the effects of acute asphyxia on cerebral energy metabolism, particularly on the dynamic changes of the concentrations of both high energy phosphates and glycolitic intermediates. These studies were performed in normoxemic fetal guinea pigs near term and demonstrate the effects of asphyxia on the fetal brain under near-optimal conditions, i.e., when fetal oxygen delivery to the cerebrum is normal and when the defense mechanisms are mature, including those based on the sympathetic nervous system. The third part describes the circulatory and the sympathetic neurohormonal responses of mature, immature, and very immature chronically prepared fetal sheep. They were tested at 0.9, 0.75, and 0.6 of the term of pregnancy. The emphasis in this section is placed on the importance of the maturation of the sympathetic nervous system during intrauterine development for a fully effective circulatory centralization during episodes of asphyxia. The last part of this review deals with metabolic changes during acute asphyxia, and stresses the importance of the sympathetic nervous system for an effective "metabolic centralization." In particular, the cellular mechanisms will be discussed by which the consumption of oxygen is reduced during asphyxia to maintain the delivery of these principal substrates to the fetal brain, thus ensuring intact survival when oxygen is in short supply.

The Incidence of Cerebral Hemorrhage in 5301 neonates Screened by Cranial Ultrasound

In recent years perinatal mortality has been reduced to such an extent that its value as an indicator of the quality of obstetrical and pediatric care is questionable. Therefore, we focused our interest on fetal and neonatal morbidity, i.e., on the degree of asphyxia during parturition, the fetal outcome, the degree of brain damage, and the psychomotor development later during life. To assess the incidence and the degree of early brain damage, the Department of Obstetrics and Gynaecology and the Department of Pediatric Radiology at the University of Giessen launched a Neonatal Cranial Ultrasound Screening program. The results were then related to obstetrical risk factors (Jensen et al. 1989a; Jensen et al., 1992; A. Jensen, V. Klingmüller, S. Sefkow, S. Bender, unpublished work).

The neonatal brain was examined by ultrasound, using the fontanellae as windows to enter the skull. The standard examination included coronary and sagittal scans, which allowed for detection of hemorrhages and other abnormalities. Modifying the classification proposed by Papile et al. (1978), who originally described four different degrees of severity for computed tomography examination, we have chosen to distinguish three degrees of severity (I–III) of cerebral hemorrhage by collapsing the two mildest degrees of cerebral hemorrhage into one category (Jensen and Schumacher 1987). Our classification is as follows:

- Peri/intraventricular hemorrhage I (PIVH I): subependymal hemorrhage with or without intraventricular hemorrhage, but without dilatation of the ventricles
- Peri/intraventricular hemorrhage II (PIVH II): like PIVH I, but with additional ventricular dilatation
- Peri/intraventricular hemorrhage III (PIVH III): like PIVH II, but with marked hemorrhage into the parenchyma of the brain

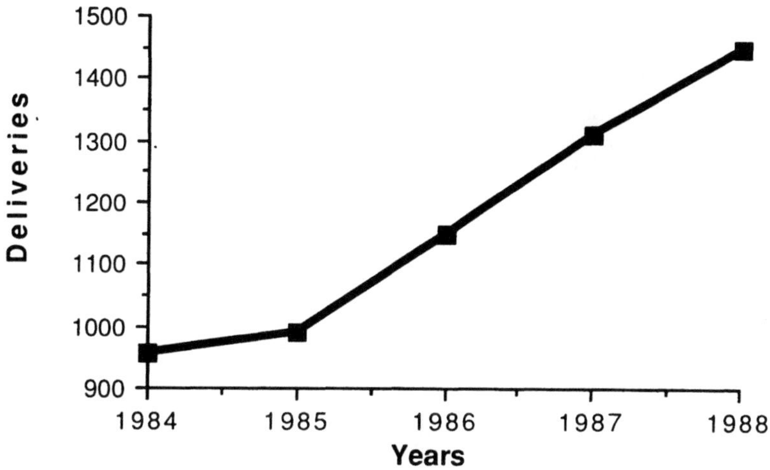

Fig. 1. The number of deliveries at the Department of Obstetrics and Gynecology in Giessen from 1984 to 1988

Between 1984 and 1988, 5853 babies were born in our obstetrical unit (Fig. 1). Of these, 5479 (94%) were examined at day 5,6 ± 3,4 (SD), using a Siemens and/or Acuson sector scanner (3.5 Hz). In babies that were not transferred to the Department of Pediatrics, the examination was performed on the day of discharge. The others were examined on arrival at the Department of Pediatrics. Those infants presenting hemorrhages were followed up dayly if appropriate. Only photo-documented ultrasound examinations entered further analysis (n = 5301). To exclude interexaminer variance, all abnormalities, including hemorrhages, were reviewed by an experienced expert.

From 5301 neonates finally entering the study, ultrasound revealed cerebral hemorrhages in 189 neonates (3.5%). Of these, 101 (1.8%) were PIVH I, 51 (0.9%) were PIVH II, and 37 (0.7%) were PIVH III. The incidence of cerebral hemorrhage was closely related to the gestational age of the neonates. This is demonstrated in Fig. 2, in which the incidence and the severity of cerebral hemorrhage is shown for five gestational age groups. There was a steep rise in both incidence and severity of hemorrhage below 34 weeks, which was even stronger below 30 weeks, which represent 75% of the term of pregnancy. In the most immature group of babies born at or below 30 weeks, about 50% had moderate or severe hemorrhages.

Interestingly, neither the fetal heart rate pattern nor the blood gas and acid-base analysis in the umbilical artery correlated well with the incidence of cerebral hemorrhage after birth. By far the best predictor of cerebral hemorrhage was the Apgar score, evaluated at 1 and 5 min post partum. When broken down into gestational age groups, it became evident that poor outcome and prematurity were strongly related to a high incidence of severe

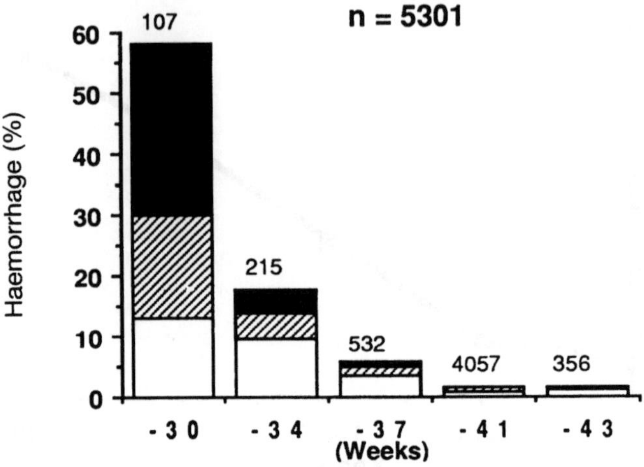

Fig. 2. Relation between cerebral hemorrhage and gestational age in human fetuses

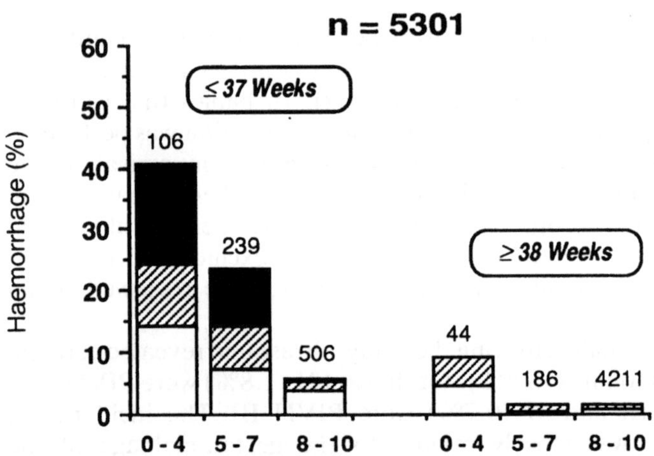

Fig. 3. Relation between cerebral hemorrhage, gestational age, and apgar score in human fetuses

hemorrhages (Fig. 3). On the other hand, the analysis also showed that neonates born near term were fairly resistant to the adverse effects of asphyxia. Although some of these were born with Apgar scores as low as 5–7 points at 1 min, their overall incidence of cerebral hemorrhage was not significantly different from those born with 8–10 points (Fig. 3). Only when they scored less than 4 points did the incidence of hemorrhage rise to about

10%. In the younger age groups the overal incidence was much higher, and hemorrhages were more severe. However, a very striking result was that – even in the group of newborns at or under 30 weeks of gestation (not illustrated) – severe hemorrhages were virtually absent if the baby was born in a good clinical condition, scoring 8 points or higher. Thus, if an asphyxic circulatory shock of the infant can be avoided by appropriate clinical measures, the effects of prematurity on both the incidence and severity of cerebral hemorrhage become less prominent.

The Giessen Neonatal Cranial Ultrasound Screening program has improved our obstetrical and neonatal care in many ways, including by identifying risk factors, improving risk management during both pregnancy and delivery, evaluating obstetrical prognostic indices such as CTG, pH, and Apgar score; evaluating the risk of obstetrical maneuvers, improving the consultation of parents, ensuring continuous feed-back on the fetal outcome and quality control of obstetrical and neonatal care, and identifying clinically important questions that require priority in basic research. All these factors along with a joint risk-orientated effort of the referring doctors, obstetricians, neonatologists and their staff have contributed to the fact that the incidence of moderate and severe cerebral hemorrhages in our neonates has declined steadily over the years (Fig. 4).

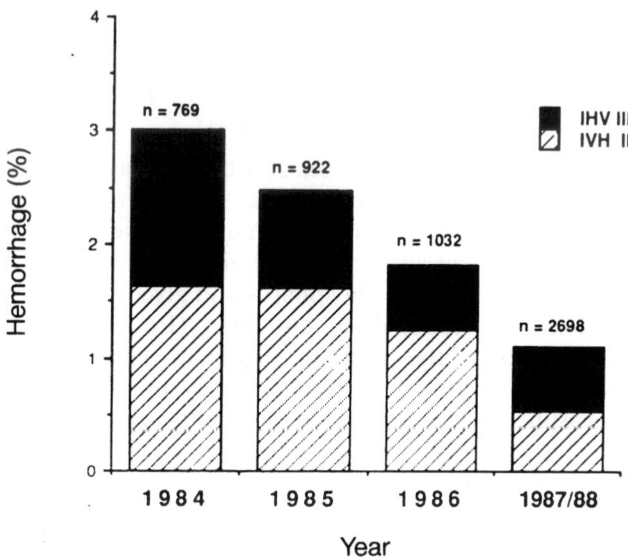

Fig. 4. The incidence of intra/periventricular hemorrhage II–III at the Department of Obstetrics and Gynecology in Giessen from 1984 to 1988

The Effects of Acute Asphyxia on High-energy Phosphates and Glycolytic Intermediates of the Fetal Cerebrum

There are numerous reports in the literature that blood flow to the fetal brain is increased when oxygen is at short supply, but this is not always true (Cohn et al. 1974; Peeters et al. 1979; Ashwal et al. 1980, 1981). We reported in 1987 for the first time that brain blood flow does not increase during acute asphyxia caused by arrest of uterine blood flow (Jensen et al. 1987a, b). Since reduction in uterine blood flow is frequent during the second stage of human labor, our model produced a reduction in oxygen delivery to the fetus by a mechanism that may be much more relevant to the clinical situation than that used in most of the other studies, i.e., maternal hypoxemia (Cohn et al. 1974; Peeters et al. 1979; Ashwal et al. 1980, 1981). According to our experimental data derived from unanesthetized fetal sheep in utero, there is a redistribution of blood flow within the brain during acute asphyxia, which favors brain stem areas at the expense of both cerebrum and choroid plexus. This would suggest that oxygen delivery to the cerebrum is not maintained under these very acute conditions and that, consequently, cerebral metabolism becomes anaerobic, lactate accumulates, and concentrations of high-energy phosphates change.

To test this hypothesis, we developed a guinea pig model (Dr. Berger) that allowed us to measure the dynamic changes of high-energy phosphates and those of glycolytic intermediates in the fetal cortex. The measurements were taken at 2-min intervals during arrest of uterine blood flow in the absence of anesthesia (Berger et al. 1990; 1991).

We studied 55 fetuses from 22 albino guinea pigs at 57 ± 2 days of gestation (term is at 68 days), before and during acute asphyxia caused by arrest in uterine blood flow. Twenty-seven of these were used to determine changes in both blood gases and acid-base balance, while 28 other fetuses were used to measure high-energy phosphates and glycolytic intermediates. The methods used are reported in detail elsewhere (Berger et al. 1991). Briefly, 10 s after decapitation of the dam, the first fetus (control) was delivered by cesarean section while the uterine artery was clamped (see Fig. 5). Both the second (2 min asphyxia) and the third fetus (4 min asphyxia) were delivered, 2 and 4 min later, respectively. In the first group ($n = 27$), mixed arterio-venous blood gases were obtained after immediate decapitation. In the second group ($n = 28$), the fetuses were instantly brought into liquid nitrogen and then stored at −80°C. Specimens from the fetal cerebral cortex (maximal thickness 1 mm) were obtained using spectrophotometry to measure high-energy phosphates and glycolytic intermediates. This test was performed in the Department of Pharmacology of the University of Marburg. The following compounds were measured: adenosine triphosphate (ATP), adenosine diphosphate (ADP), adenosine monophosphate (AMP), creatine phosphate (CP), glucose, lactate, fructose-1,6-bisphosphate (FDP), and pyruvate. The results are given as means ± SEM (Standard error of mean). For statistical

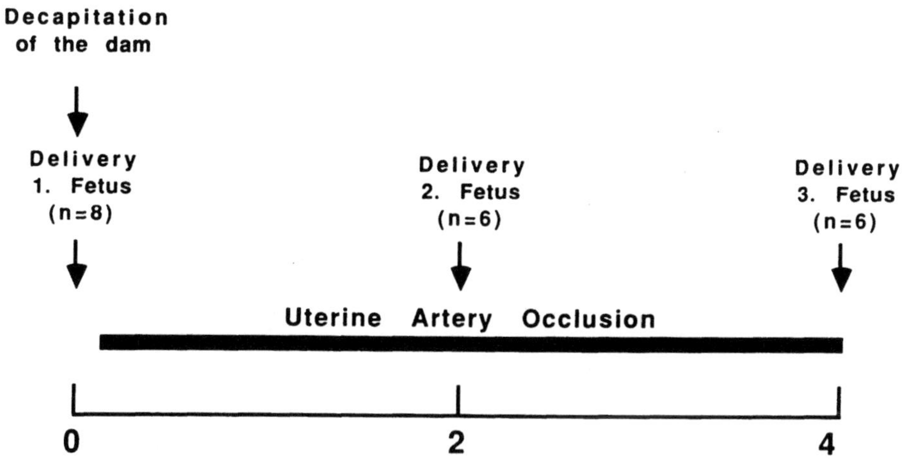

Fig. 5. Experimental protocol to study high-energy phosphates in the fetal brain. Guinea pigs near term

evaluation, the Wilcoxon rank test and linear regression analysis were used.

Blood gas values and acid-base variables of the control groups were within the normal range (Fig. 6) and comparable to those reported for chronically instrumented fetal sheep near term (Iwamoto et al. 1983; Jensen and Lang 1988). During 2 and 4 min asphyxia, there was a progressive decrease in mixed arterio venous pO_2 and pH (Fig. 6).

Control values for high-energy phosphates, glucose, and lactate in the fetal brains (Figs. 6, 7) were comparable to those reported for normoxemic fetal and adult rats (Siesjö 1981). During 2 and 4 min asphyxia there was a progressive fall in cerebral concentrations of ATP, CP, and glucose, and an increase in those of ADP, AMP and lactate (Figs. 6, 7).

These findings in fetal guinea pigs near term may be clinically important, because during human labor, acute asphyxia caused by reduction in uterine blood flow is fairly frequent. The findings demonstrate for the first time that brief episodes of arrest of uterine blood flow, e.g., for 2 min, are accompanied by a degradation of high-energy phosphates and by lactate accumulation in the fetal cerebral cortex. Secondly, these results confirm our hypothesis derived from the fetal sheep model that oxygen delivery to the cerebrum decreases during acute asphyxia (Jensen et al. 1987a, b). Since this is accompanied by anaerobic metabolism, as evidenced by increasing lactate and decreasing glucose concentrations in the cerebrum, these changes may eventually lead to brain damage and neuronal loss if asphyxia is prolonged (Myers and Yamaguchi 1976). However, if asphyxia is brief and is followed by an appropriate recovery period, it is likely that the fetal brain can sustain a number of repeated asphyxial episodes because blood flow to the fetal brain

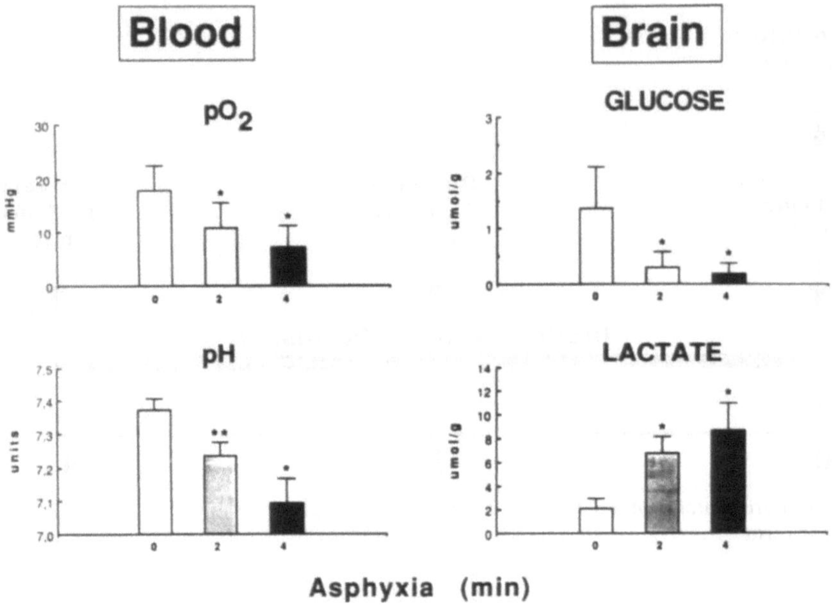

Fig. 6. Changes in blood and brain values during acute asphyxia. Guinea pigs near term

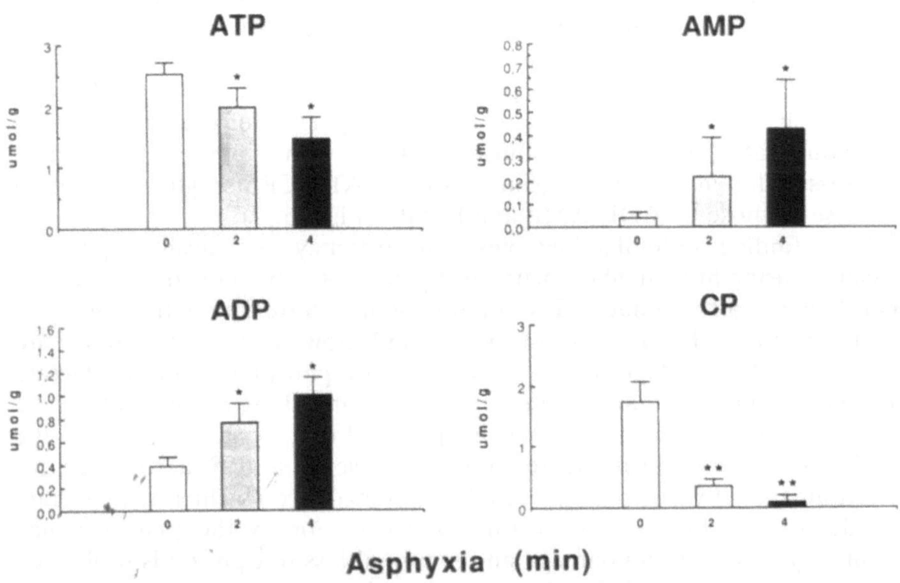

Fig. 7. Energy-rich phosphates of the brain during acute asphyxia. Guinea pigs near term

increases two- to threefold after acute asphyxia (Jensen and Lang 1988, 1989). Thus, the oxygen debt of the brain is probably compensated in large part by intermittent cerebral hyperperfusion. The mechanism by which cerebral blood flow is increased after asphyxia is not known, but adenosine, the endproduct of the degradation of ATP, may be involved because of its well-known vasodilatory properties.

It seems important to notice that the changes in the concentrations of high-energy phosphates and glycolytic intermediates during asphyxia occurred in spite of the fact that these fetuses were both normoxemic before asphyxia and mature as far as the sympathetic nervous system is concerned (Jelinek and Jensen 1988, 1991). Hence, fetal capabilities to counter the adverse effects of asphyxia on the brain were near optimal, because important defense mechanisms are already operative at this point in development. The first line of defense in this regard is the circulatory redistribution towards the brain and heart to maintain oxygen delivery during asphyxia. This mechanism is largely dependent on the maturity of the sympathetics, as we will see in the next part of this review. The second line of defense to ensure intact survival of the fetus includes all of the mechanisms conserving substrates, e.g., those which reduce consumption of oxygen and glucose in peripheral organs. These defense mechanisms act on a cellular level and are in part dependent on the maturity of the sympathetic nervous system, as we will see in the last part of this review.

Circulatory and Sympathetic Responses to Acute Asphyxia at 0.9, 0.75, and 0.6 of Term

Previous studies on the effects of asphyxia in human infants (Jensen and Künzel 1980; Paulick et al. 1985) and in fetal sheep have shown that *near term* fetal organ blood flow is redistributed in favor of the brain and the heart, at the expense of peripheral organs, to maintain oxygen delivery to vital organs (Campbell et al. 1967; Cohn et al. 1974; Ashwal et al. 1981; Jensen et al. 1987a, c, Jensen and Berger 1991). Catecholamine concentrations have been shown to rise steeply during asphyxia (Jensen et al. 1987b), and ablation of the sympathetics by various methods has been shown to blunt a number of the circulatory responses of the fetus during asphyxia (Iwamoto et al. 1983; Jensen and Lang 1988). Therefore, it has previously been proposed that the sympathetic nervous system is essential for an effective circulatory centralization that ensures intact survival of the fetus (Jensen and Lang 1988; Jensen et al. 1989b). However, there is no information on the changes of the circulatory and sympathetic responses to acute asphyxia during development when the sympathetic nervous system is still immature; its future study may be of major importance for understanding the mechanisms which are responsible for the fact that premature babies suffer much more frequently from brain damage in the perinatal period than do mature babies.

From the Giessen Cranial Ultrasound Screening program, it is evident that the largest increase in both the incidence and severity of cerebral hemorrhage, relative to that at term, happens to be at around 0.75 of term, i.e., at 30 weeks (see Fig. 2). We felt that this fact required investigation, because factors other than the known structural peculiarities of the fetal brain could be involved, e.g., those related to insufficient circulatory centralization. Furthermore, it has become clinical practice in some obstetrical units to deliver premature babies as early as at 24–25 weeks of gestation, i.e., at 0.6 of term, when fetuses are barely viable in an extrauterine environment and when the risks of both brain damage and neurologic sequelae later during life are high. We therefore wished to determine the fetal cardiovascular and catecholamine responses before, during, and after acute asphyxia at various stages of development equivalent to the 37th, 30th, and the 24–25th week of human gestation. This would allow us to assess possible interrelations between immature circulatory responses and immature sympathetic responses that might be associated with higher incidences of brain damage in premature fetuses.

We studied 17 chronically prepared fetal sheep at 0.9, 0.75, and 0.6 of term, i.e., at 130 ± 2 ("mature", $n = 5$), 110 ± 1 ("immature", $n = 7$), and 90 ± 3 ("very immature", $n = 5$) days gestation (term is at 147 days). The studies took place before, during, and after asphyxia caused by arrest of uterine blood flow for 2 min. The methods used for catheter placement, maintenance of the sheep, blood flow measurements using sequential injections of different batches of isotope-labeled microspheres (15 μm diameter) at minute-intervals, and withdrawal of blood to determine blood gases, acid-base status, and catecholamine concentrations have been described elsewhere (Jensen et al. 1987a–c; Jensen and Lang 1988). The concentrations of catecholamines in the fetal plasma, e.g., epinephrine and norepinephrine, were determined by high-performance liquid chromatography (HPLC) with electrochemical detection (ECD) as described by Jelinek and Jensen (1988). The experimental protocol was as follows: after a control period, preductal blood samples were withdrawn, and the first batch of microspheres was injected into the inferior vena cava (at −1 min), while reference blood samples were taken from both the brachial and femoral arteries. Uterine blood flow was then arrested (at 0 min) for 2 min by occlusion of the descending maternal aorta using a snare (Jensen et al. 1987a–c; Jensen and Lang 1988). All measurements were repeated at 1, 2, 3, 4, and 30 min. During the whole experiment, maternal (below occlusion) and fetal arterial blood pressure, fetal heart rate, and amniotic fluid pressure were recorded continuously on a polygraph (Hellige, Freiburg, FRG). After the experiment, both fetus and mother were killed using pentobarbitone and saturated potassium chloride, administered intravenously. The fetus then was perfused with formalin and dissected to obtain organ specimens for blood-flow measurements as previously described (Jensen et al. 1987a, b). The results are given as mean ± SEM, and statistical analysis was performed by multivariate analysis of variance in a repeated measures design.

In the control period, blood gases and acid-base variables were within the normal range for chronically prepared fetal sheep, and there were no significant differences between groups. However, arterial blood pressure was significantly lower (Fig. 8), and heart rate was significantly higher (Fig. 9) in

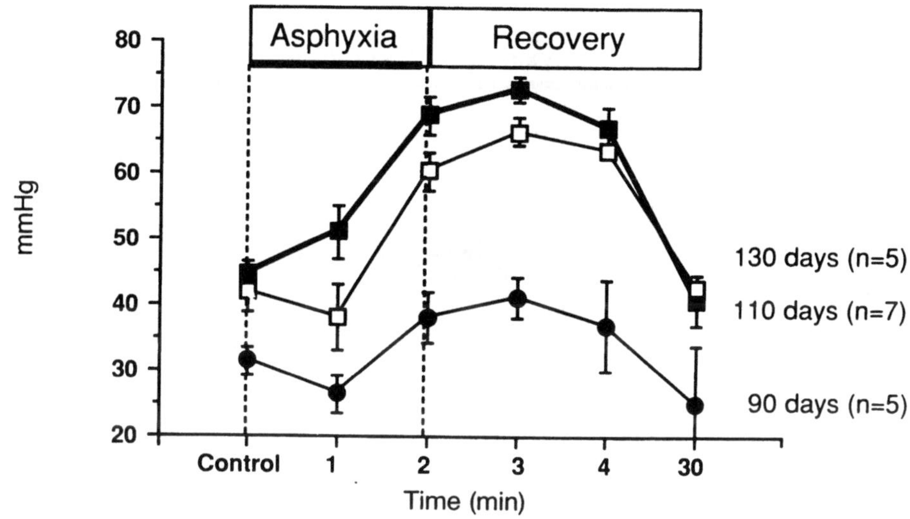

Fig. 8. Arterial blood pressure of fetal sheep during asphyxia at 0.6, 0.75, and 0.9 of term

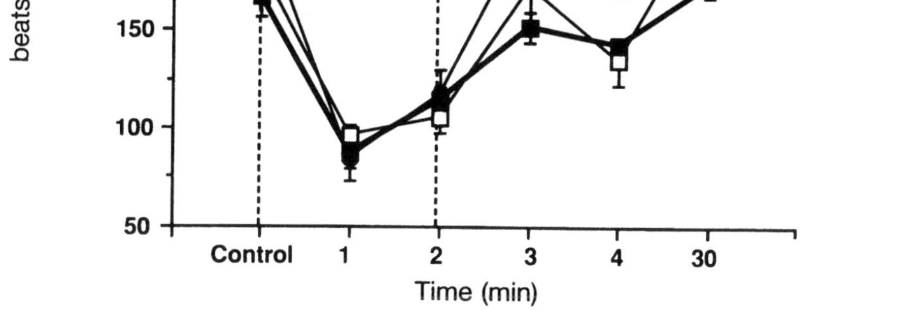

Fig. 9. Heart rate of fetal sheep during asphyxia, at 0.6, 0.75, and 0.9 of term

very immature fetuses (90 days), than in both immature (110 days) and mature (130 days) fetuses. Blood flow to the heart (left cardiac ventricle) was similar (Fig. 10), but blood flow to the cerebrum was different between groups in an age-related manner (Fg. 11). Cerebral blood flow was highest in mature, lower in immature, and lowest in very immature fetuses, whereas that

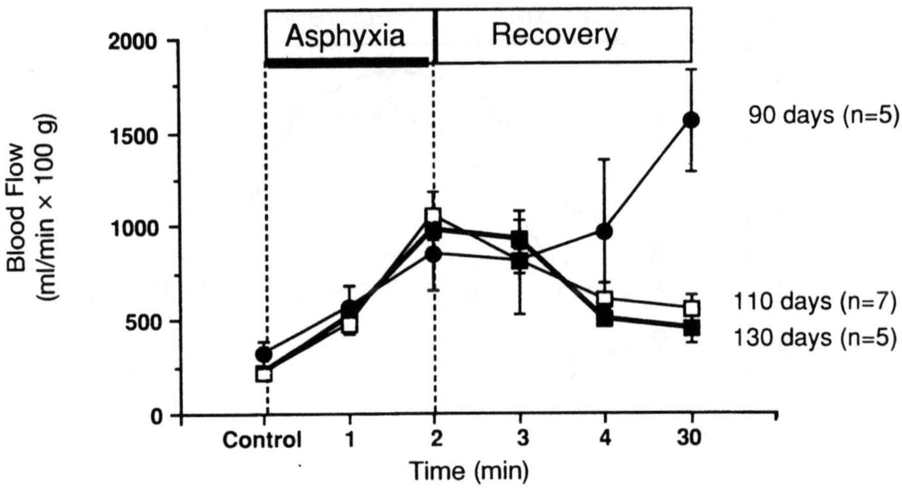

Fig. 10. Left cardiac ventricle blood flow of fetal sheep during asphyxia at 0.6, 0.75, and 0.9 of term

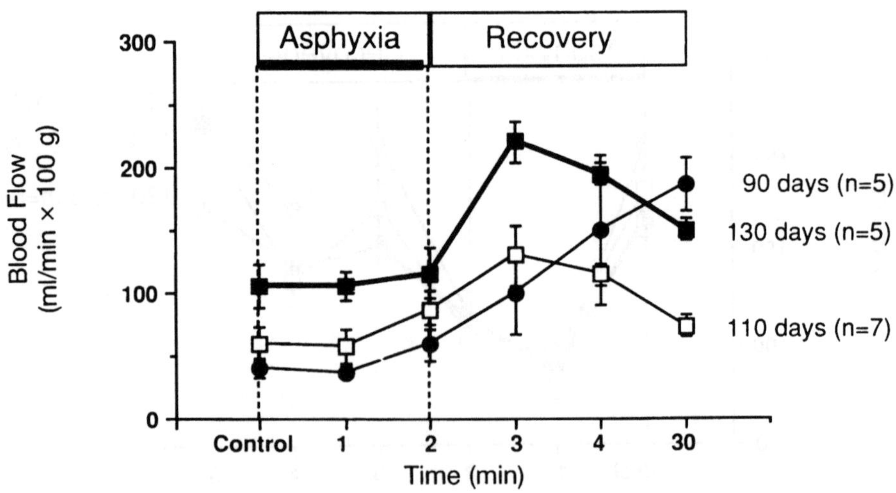

Fig. 11. Cerebral blood flow of fetal sheep during asphyxia at 0.6, 0.75, and 0.9 of term

to the germinal matrix (Fig. 12) was not different in immature and very immature fetuses (there is no germinal matrix in mature fetuses). Blood flow to the carcass (Fig. 13) was similar in the two older groups, but higher in the very immature fetuses.

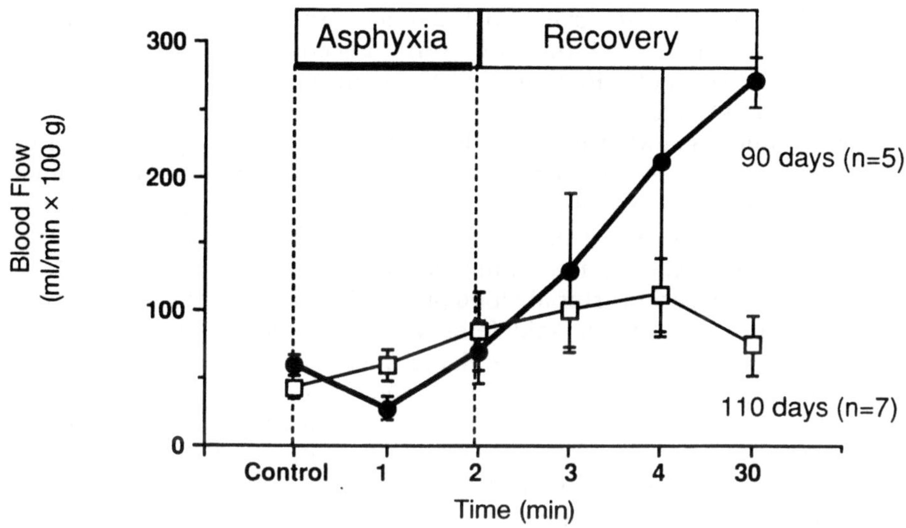

Fig. 12. Germinal matrix blood flow of fetal sheep during asphyxia at 0.6 and 0.75 of term

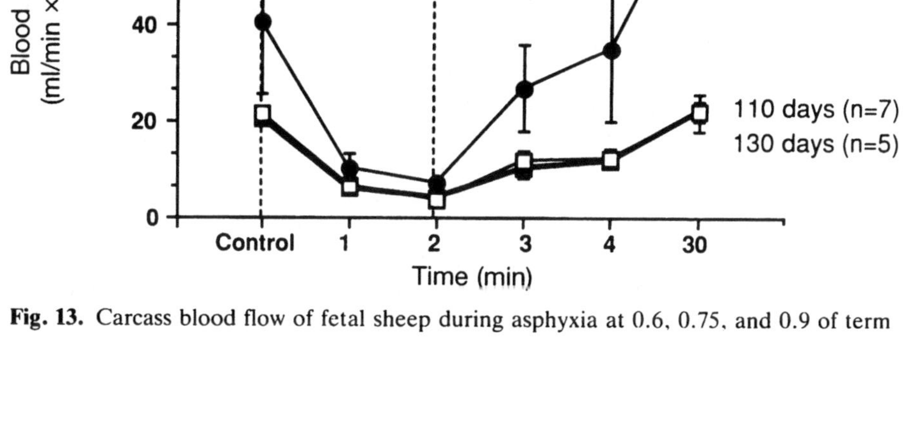

Fig. 13. Carcass blood flow of fetal sheep during asphyxia at 0.6, 0.75, and 0.9 of term

During asphyxia there was a progressive fall in arterial pO_2, oxygen saturation (SaO_2), and pH in all groups. There were no significant differences in fetal heart rate (Fig. 9) between groups, but the arterial blood pressure response (Fig. 8) was very different. In mature fetuses, arterial blood pressure rose progressively during asphyxia, whereas it fell transiently at 1 min in both immature and very immature fetuses, resulting in significant differences between the groups at 2 min of asphyxia, even though blood pressure increased also in the latter groups at 1–2 min of asphyxia. These age-related changes in arterial blood pressure were reflected by those in plasma catecholamine concentrations, in that the concentrations of both norepinephrine (Fig. 14) and epinephrine (Fig. 15) during asphyxia were highest in mature fetuses, somewhat lower in immature fetuses, and by far lowest in very immature fetuses. Blood flow to both the cerebrum and the germinal matrix did not change significantly during asphyxia (Figs. 11, 12), whereas that to the heart increased progressively with no differences between groups, and that to the carcass decreased invariably in all groups (Figs. 10, 13).

In the recovery period arterial blood pressure fell to control values or even below as observed in 90-day-old fetuses (Fig. 8), whereas the heart rate returned to normal (Fig. 9). These changes were accompanied by a fall in catecholamine concentrations to control values (Figs. 14, 15). Blood flow to the cerebrum returned to normal in mature and immature fetuses, not quite reaching control values, whereas in very immature fetuses blood flow increased progressively to 300% at 30 min recovery (Fig. 11). Even more surprising was the rise in blood flow to the germinal matrix in the very immature fetuses, as compared to the immature (110 days) fetuses, which peaked at 30 min recovery, reaching almost 500% of the control value (Fig. 12). An almost similar rise in recovery blood flow was measured in the heart (+ 400%) of the very immature fetuses, whereas in the older fetuses myocardial blood flow decreased, not quite reaching control values (Fig. 10). Carcass blood flow in the older fetuses recovered to control values, whereas that of the very immature fetuses increased 100% above control values at 30 min (Fig. 13).

With respect to chronically prepared fetal sheep, these results show that the circulatory and sympathetic responses to acute asphyxia change considerably in an age-related manner during stages of development equivalent to the 37th, 30th, and 24–25th week of human gestation. To our knowledge this has not been reported previously. The most obvious correlation in age-related differences was between arterial blood pressure and plasma catecholamine concentrations, in that a "mature" blood pressure response, i.e., progressive increase in blood pressure during asphyxia, was shown to be dependent on massive rises in concentrations of both norepinephrine and epinephrine in the fetal plasma. This response was only achieved by 130-day-old fetuses. Immature fetuses at 110 days of gestation reached only 80% and 50% of the peak concentrations of norepinephrine and epinephrine in mature fetuses, and blood pressure failed to rise after 1 min of asphyxia. They also did not reach the values of mature fetuses thereafter. Even more obvious was the

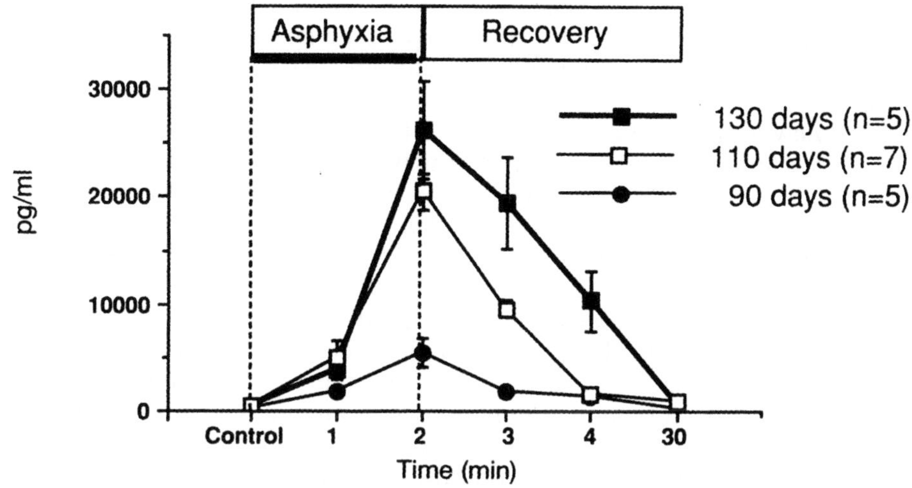

Fig. 14. Norepinephrine concentration during asphyxia of fetal sheep at 0.6, 0.75, and 0.9 of term

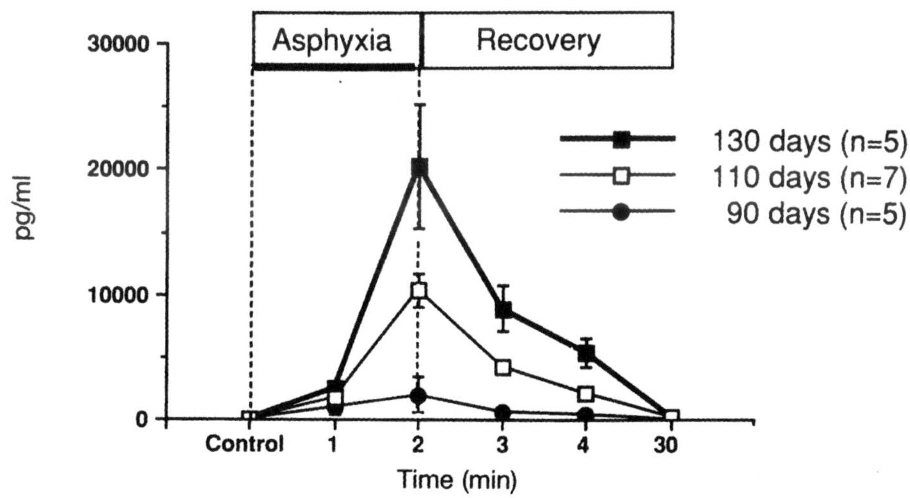

Fig. 15. Epinephrine concentration during asphyxia of fetal sheep at 0.6, 0.75, and 0.9 of term

correlation between arterial blood pressure and catecholamine concentrations in the 90-day-old fetuses, where both pressure and sympathetic responses during asphyxia were hardly statistically different from zero. This holds particularly true for the change in epinephrine concentrations. These developmental changes in pressure and catecholamine responses are accompanied by age-related differences in circulatory centralization of the fetuses during asphyxia, suggesting that the sympathetic nervous system plays an important part in the development of a fully effective circulatory centralization.

This view is supported by the fact that blood flow to peripheral organs, e.g., to the fetal carcass, is considerably higher before and after asphyxia and does not decrease as much as in mature fetuses. Hence, peripheral vasomotor control, which depends largely on peripheral sympathetic neurons (Jensen and Lang 1988, 1989), is virtually absent at 90 days of gestation. This lack of peripheral vascular tone, which is probably accompanied by an increased cardiac output, is also reflected by the poor arterial blood pressure and by increased blood flow to the heart during late recovery in this group.

With regard to brain damage, the most important result of our studies may be the finding that very immature fetuses show an unusual pattern of blood flow changes during and after asphyxia in the cerebrum and in the germinal matrix. First, during asphyxia, blood flow and hence oxygen delivery to the cerebrum of very immature fetuses is only 40% of that of mature fetuses. Secondly, and possibly more important with regard to the development of cerebral hemorrhage, there is a distinct postasphyxial hyperperfusion of both the cerebrum and the germinal matrix of almost 300% and 500%, respectively. Thus, there may be at least two important differences between very immature and mature fetuses. First, a given asphyxial episode produces a much more severe cerebral hypoxic insult in very immature than in mature fetuses, and secondly, in addition to this cerebral hypoxic insult, the cerebrum and particularly the germinal matrix are exposed to postasphyxial hyperperfusion.

These obvious differences in cerebral blood flow may explain in part why very immature fetuses are more susceptible to brain damage in general and to cerebral hemorrhage in particular than are mature fetuses. This hypothetical mechanism of brain damage seems particularly attractive since it is well known from clinical practice in neonatal intensive care units that very immature neonates are, as far as hemorrhage into the vulnerable germinal matrix is concerned, very sensitive to any intervention that might increase arterial blood pressure and cerebral blood flow.

The actual mechanism causing hyperperfusion of the immature brain and heart after asphyxia is not known. Yet since circulatory centralization is poor, due to insufficient sympathetic vascular control in peripheral organs e.g., the carcass, it is likely that oxygen delivery to and hence oxygen consumption by peripheral organs are high in contrast to mature fetuses (Jensen et al. 1987c). This assumption is based on previous findings both in vivo (Acheson et al. 1957; Acker and Lübbers 1977; Jensen et al. 1987a–c) and in vitro (Braems and Jensen 1990), which showed that oxygen consumption of a given tissue depends largely on oxygen delivery and oxygen extraction. Thus, it seems reasonable to conclude, that in very immature fetuses, i.e., at about 0.6 of term, circulatory and metabolic centralization during and after asphyxia are ineffective, and that the immaturity of the sympathetic nervous system may account for such findings. Conversely, it also seems reasonable to conclude that the maturity of the sympathetic nervous system near term may account in part (notwithstanding the known structural changes in the brain) for the relative resistance of the fetus to the more adverse effects of asphyxia on the fetal brain. Therefore, we believe that the sympathetic nervous system plays an important part in preventing brain damage during asphyxia in the perinatal period (Jensen and Berger 1991).

The role of the Sympathetic Nervous System in Controlling Fetal Oxygen Consumption During Acute Asphyxia

Effects of Arrest of Uterine Blood Flow

While there is no doubt that the sympathetics are essential for an effective circulatory centralization and hence for the maintenance of cerebral blood flow during acute asphyxia, their role in controlling fetal metabolism is less clear. We therefore investigated this matter in unanesthetized fetal sheep near term, when the sympathetic nervous system is mature. In particular, we were interested in the time course of the changes in oxygen consumption and organ blood flow during acute asphyxia, and in how these dynamic changes relate to changes in plasma catecholamine concentrations. We developed a model that allowed us both to measure oxygen consumption directly and to determine the distribution of blood flow before and during acute asphyxia at 1-min intervals.

The methods used have been reported in detail elsewhere (Jensen et al. 1987c). Briefly, nine unanesthetized fetal sheep weighing 4.6 ± 0.5 kg at a mean gestational age of 130 ± 2 days (term is at 147 days) were studied in utero in five chronic and four acute preparations before and during asphyxia, caused by arrest of uterine and ovarian blood flow. This was achieved by the insertion of an inflatable balloon catheter around the descending maternal aorta.

To assess the time course of changes in total oxygen consumption of the fetus, placenta, and membranes, the decrease (Δ) of the oxygen content (ml/dl), i.e., the fractional O_2 saturation $\times$ hemoglobin concentration (g/dl) $\times$ 1.34 (dissolved O_2 is negligible), in the blood of the descending aorta was determined at 1-min intervals (e.g., $\Delta O_2/min_1$ = O_2 content at 1 min of asphyxia) and multiplied by the estimated fetal blood volume (dl/kg) (Creasy et al. 1970). Values are expressed as ml O_2/min per kg fetus. To validate the procedure, a healthy, chronically prepared fetus near term was used to sample blood at 1-min intervals simultaneously from a femoral and a carotid artery during a 4-min period of aortic occlusion. The calculated decrease (ΔO_2) of the arterial O_2 content in the two vessels after 1, 2, 3, and 4 min correlated well ($r = 0.993$), suggesting that the error is small when femoral instead of carotid arterial blood is used to calculate total oxygen consumption directly (Barcroft et al. 1939).

To determine the time course of the redistribution of blood flow to the fetal organs, blood flow was measured by a modified microsphere technique (Rudolph and Heymann 1967; Heymann et al. 1977). Isotope-labeled microspheres (^{141}Ce, ^{113}Sn, ^{103}Ru, ^{95}NB, and ^{46}Sc, 16 µm diameter) were injected into the inferior vena cava during the control period and at 1, 2, 3, and 4 min during asphyxia. Because we wanted to measure blood flow at short intervals, we modified the original microsphere method developed by Rudolph and Heymann (1967), in that we injected the batches of differently labeled microspheres sequentially during continuous withdrawal of reference blood over 4.5 min, the volume being simultaneously replaced by maternal blood maintained at 39 °C in a waterbath. Thus, the methodological error due

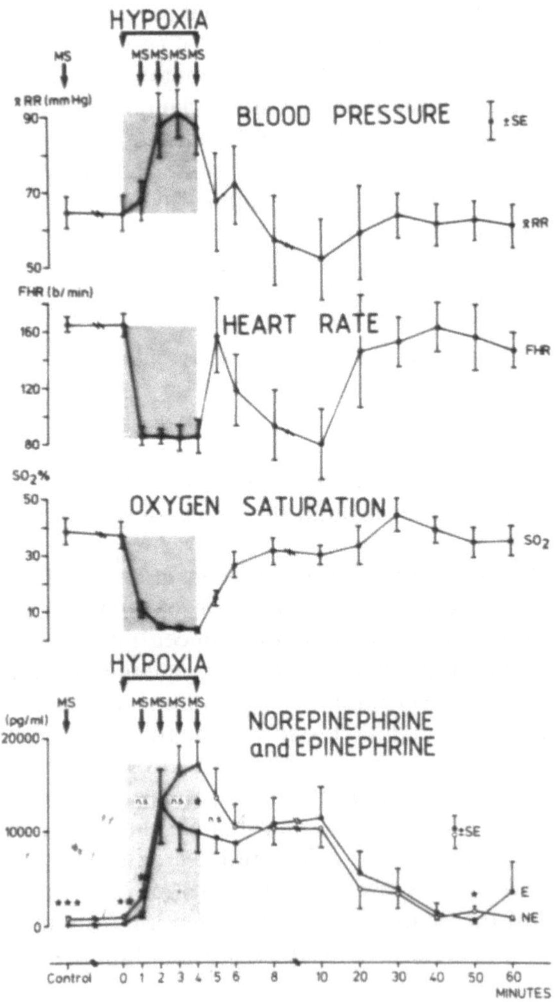

Fig. 16. Arrest of uterine blood flow. Experimental protocol. Changes in fetal physiologic variables in sheep near term

to the variance of four separate reference sampling procedures was reduced to a quarter, and observations of acute blood flow changes at intervals of 1 min were possible. Furthermore, the time of withdrawal (270 s vs. 360 s) and hence the volume exchanged were reduced by 25%. This advantage is of increasing importance in smaller animals and in proportion to both the number of reference samples and the rate of withdrawal.

Arrest of uterine blood flow for 4 min caused severe fetal asphyxia; heart rate, arterial O_2 saturation of hemoglobin, O_2 content, and pH decreased and arterial blood pressure, pCO_2, lactate, and norepinephrine and epinephrine concentrations increased (Fig. 16). The fall of the mean arterial O_2 content was paralleled by an exponential fall of the total O_2 consumption of fetus, placenta, and membranes with time (Fig. 17). There was a close linear correlation between these two variables after each minute of asphyxia, but the slope of the regression line was steeper after 2 min than after 1, 3, and 4 min. During the control period, the estimated O_2 consumption (extrapolated from the regression in Fig. 17) was 9.8 ± 1.2 ml/min per kg.

Within 1 min of asphyxia there was a significant decrease in blood flow to most of the peripheral organs, including the skin, the scalp, and the choroid

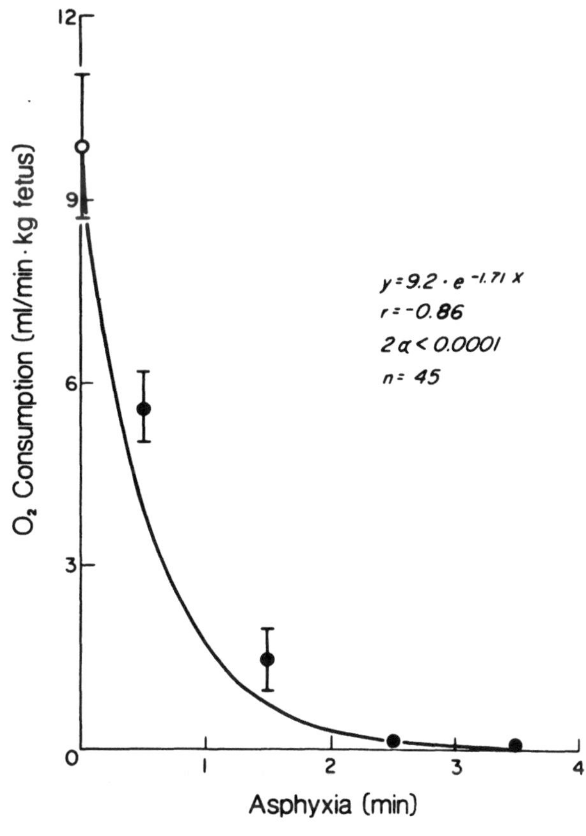

Fig. 17. Total oxygen consumption of the fetus, placenta, and membranes in sheep near term

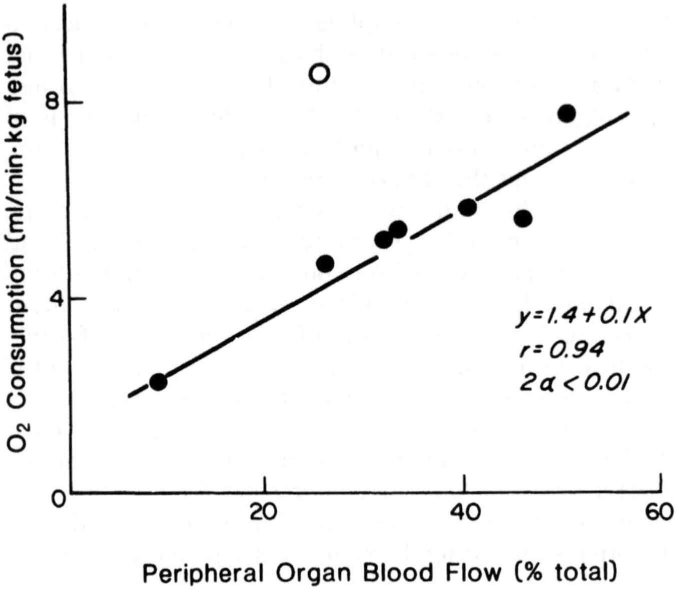

Fig. 18. Changes in oxygen consumption in relation to peripheral blood flow of sheep near term

plexus, to between 46% and less than 4% of control values. There was also a significant decrease to the cerebrum, whereas blood flow to the cardiac ventricles and brain stem were well maintained. In eight of the nine fetuses, the decrease in blood flow to peripheral organs was closely correlated with the decrease in total oxygen consumption (Fig. 18). These changes were accompanied by a steep rise in both norepinephrine and epinephrine concentrations in the fetal plasma.

After 2 min of asphyxia, plasma catecholamine concentrations continued to rise, while blood flow to peripheral organs and total oxygen consumption continued to decrease. Both blood flow and oxygen consumption reached minimal values at 3 and 4 min of asphyxia. Only in fetuses that died in the recovery period ($n = 4$) was the decrease in blood flow to peripheral organs transient, it increased after 3 min of asphyxia, suggesting circulatory decentralization in these fetuses.

These results showed that the time course of the fall in fetal oxygen consumption and in blood flow to the peripheral organs is rapid and is accompanied by steep increases in plasma catecholamine concentrations. Since arterial blood pressure rose during asphyxia, the decrease in peripheral blood flow was caused by vasoconstriction. The rapidity and uniformity of this vasoconstriction in most of the peripheral organs, along with the evidence that blood flows to the fetal organs correlates closely with increased plasma catecholamine concentrations (Jensen et al. 1987c), suggest a generalized

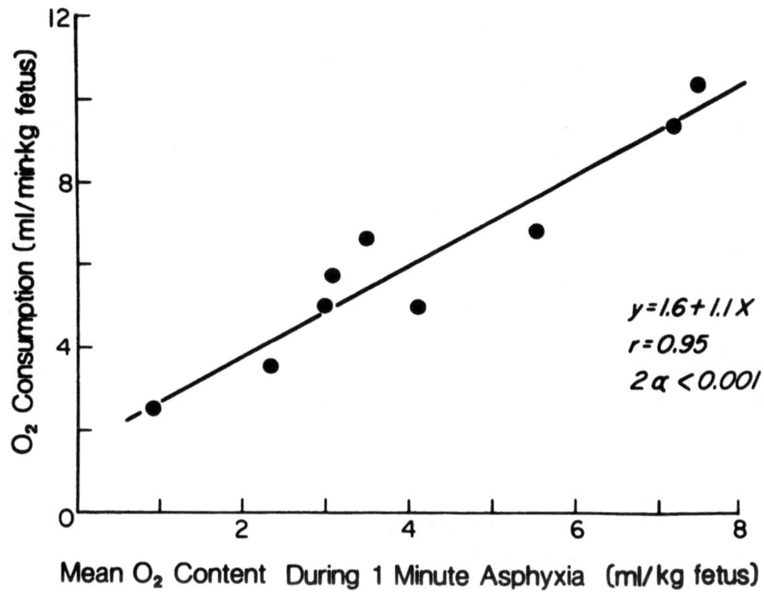

Fig. 19. Changes in oxygen consumption in relation to arterial oxygen content in sheep near term

vascular reflex mechanis involving the sympathetics (Bernthal and Schwind 1945; De Burgh, Daly and Scott 1963; Dawes et al. 1968; Killip 1963). This reflex is mediated in part through peripheral arterial chemoreceptors (Hanson and Jensen 1989, Jensen and Hanson 1989).

The similarity of the time courses of the changes in O_2 consumption, peripheral blood flow, and plasma catecholamine concentrations suggest that the mechanisms may be interrelated. All variables change as a result of a fall in arterial O_2 content (Fig. 16). It has been shown for whole cell preparations (Wilson et al. 1979), various organs (Acker and Lübbers 1977; Bristow et al. 1983), and for the entire conceptus (Acheson et al. 1957) (Fig. 19) that the amount of oxygen available determines the amount of oxygen consumed. Since the blood flow to an organ determines the amount of oxygen delivered to that organ, we suggest that changes in vasomotor activity, e.g., caused by the sympathetic nervous system, play an important part in regulating fetal O_2 consumption by changing blood flow and hence O_2 delivery (i.e., blood flow $\times$ O_2 content) to peripheral organs. Whether the fall in placental O_2 consumption parallels that of the fetus when maternal and/or fetal O_2 delivery to the placenta is decreased (Campbell et al. 1966) and how that fall in placental O_2 consumption might affect its endocrine functions, remains to be established.

Again, the key role of oxygen as a determinant of O_2 consumption is demonstrated by the close correlation between oxygen content and oxygen consumption (Fig. 19). It is also supported by the observation, made in one fetus, that after pretreatment with oxygen the total oxygen consumption was high even though peripheral blood flow was poor (Fig. 18).

In summary, we concluded from these results that during acute asphyxia the sympathetic nervous system is involved in the regulation of fetal oxygen consumption through the following mechanism: As soon as the oxygen content in the fetal carotid artery falls (e.g., during arrest of uterine blood flow) peripheral arterial chemoreceptors (e.g., in the carotid sinus) are activated and sympathetic tone rises (Hanson and Jensen 1989). These changes, which are mediated by both the sympathoneuronal and the sympathomedullary system, result – together with increasing catecholamine concentrations – in a fall in blood flow to peripheral organs and in an increase in arterial blood pressure. Consequently, oxygen delivery to peripheral organs decreases and the extraction of oxygen by the tissues rises to a maximum. Beyond this maximal oxygen extraction, as assessed by an increased arteriovenous difference across a given organ or tissue, the tissue becomes hypoxic and oxygen consumption decreases. Through this mechanism the sympathetics reduce peripheral oxygen consumption and total fetal oxygen content is preserved to maintain oxydative metabolism in the fetal heart and brain. Thus, the sympathetic nervous system is involved in both circulatory and metabolic centralization during acute asphyxia to ensure intact survival of the fetus (Jensen et al. 1987c; Jensen and Lang 1988).

Effects of Graded Reduction in Uterine Blood Flow

These conclusions were confirmed by a study performed in Dr. Rudolph's laboratories at the Cardiovascular Research Institute, University of California, San Francisco, USA where produced graded reduction in uterine blood flow, rather than total arrest of uterine blood flow, in nine chronically prepared fetal sheep near term (Jensen et al. 1991). We reduced fetal oxygen delivery by 50% in two steps and injected radiolabeled microspheres for organ blood flow measurements (Rudolph and Heymann 1967). These measurements were taken at control and at steps 1 and 2 of the reduction in uterine blood flow into the superior and inferior vena cava and into the umbilical vein simultaneously at each of the three time points (i.e., 3 $\times$ 3 batches of microspheres). Blood samples to measure blood gases, acid-base status, and glucose and lactate concentrations were withdrawn immediately before each blood flow measurement and at 30 and 60 min in the recovery period, i.e., after the snare had been released. Blood samples were taken simultaneously from a femoral and a carotid artery, inferior and superior vena cava, umbilical vein, and superior sagittal sinus ($n = 4$). Thus, among other variables not relevant to this review, we were able to calculate oxygen delivery to various organs and arteriovenous differences in oxygen content across the umbilical circulation, the upper and lower body section, and the brain. Furthermore, oxygen consumption of these organ parts was determined.

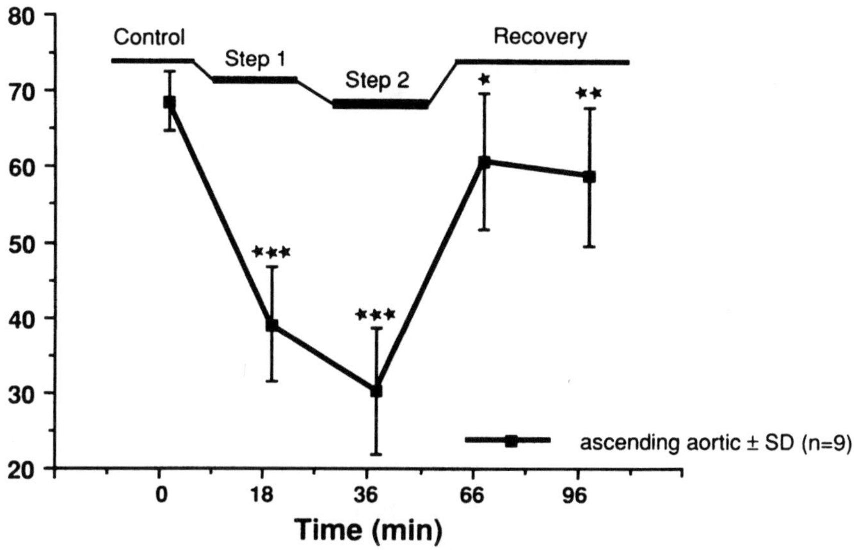

Fig. 20. Graded reduction in uterine blood flow. Experimental protocol. Changes in arterial oxygen saturation of hemoglobin in sheep near term

Graded reduction of uterine blood flow reduced total fetal oxygen delivery (i.e., umbilical venous oxygen content × umbilical blood flow) by 32% and 50% at steps 1 and 2, respectively. This resulted in a progressive decrease in oxygen saturation of hemoglobin and in oxygen content in blood obtained from each fetal vessel (Fig. 20), a marked increase in pCO_2 from 47 ± 4 to 60.0 ± 7 mmHg for the carotid artery (Fig. 21), and a decrease in pH to 7.25 ± 0.08. The oxygen content difference across the placenta (umbilical vein–descending aorta) did not change significantly during step 1, but decreased significantly during step 2 (from 3.0 ± 0.9 to 1.7 ± 0.9 ml/dl, $p < 0.001$). Interestingly, the difference in arteriovenous oxygen content across the upper body (ascending aorta–superior vena cava) decreased from 2.3 to 0.9 ml/dl, but that across the lower body (descending aorta–abdominal inferior vena cava) did not change significantly (1.98 to 1.77 ml/dl). There was also a decrease in the difference in oxygen content between ascending and descending aortic blood from 0.93 to 0.45 ml/dl. During recovery all variables tended to normalize within 60 min, though pH was still significantly reduced.

These changes were accompanied by a circulatory centralization in favor of the brain, heart, and adrenals and at the expense of the carcass, body skin, and scalp, whereas combined ventricular output did not change and umbilical blood flow was maintained. The fall in blood flow to the lower body segment at step 2 was linearly correlated with the increase in pCO_2 in the carotid artery ($y = 320 - 3.7\,x$; $r = 0.83$; $n = 9$; $p = 0.001$). These changes were accompanied by a 44% decrease in total fetal oxygen consumption during

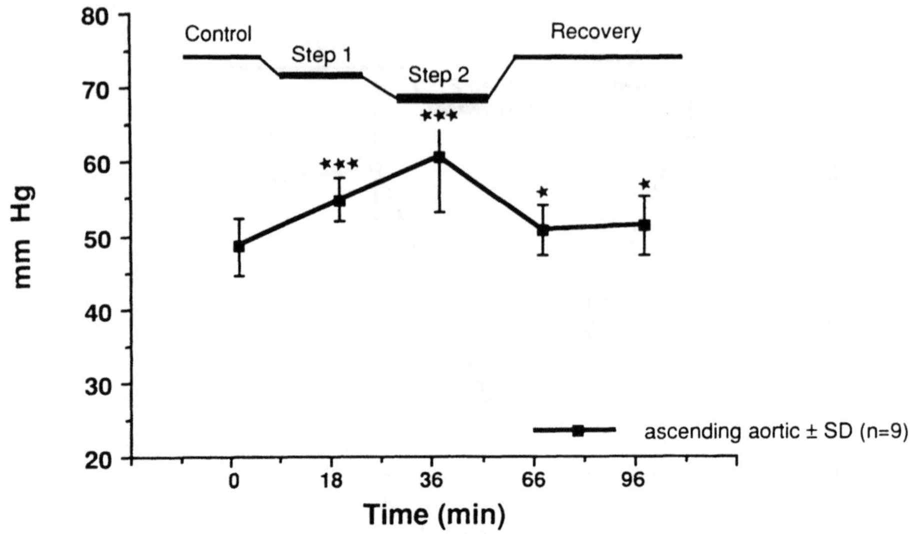

Fig. 21. Graded reduction in uterine blood flow. Experimental protocol. Changes in pCO_2 in sheep near term

step 2. Oxygen delivery to the brain and to the myocardium was maintained, but that to the adrenal glands increased to about twice control levels. Related to the decrease in both blood flow and arterial oxygen content, there was a drop in oxygen delivery to most other organs and an increase in oxygen extraction across the upper and lower body segment to 38 ± 8 and 53 ± 15, respectively; while the extraction across the brain remained constant. Interestingly, there was no increase in the arteriovenous difference in oxygen content across the lower body during reduction in uterine blood flow, suggesting that oxygen consumption fell when oxygen delivery was reduced. This was also reflected in a close linear correlation between oxygen delivery to and oxygen consumption of the upper and lower carcass and of the total lower body segment during both step 1 and step 2 of uterine blood flow reduction (Fig. 22). Oxygen consumption of the brain, as calculated from total cerebral blood flow and oxygen content difference between ascending aortic and sagittal sinus blood, did not change significantly during uterine blood flow reduction.

This study provides further evidence that during asphyxia oxygen consumption of a given part of the body, e.g., of the lower body segment, is closely related to oxygen delivery to that part. Thus, the close relationship between these two variables does not only hold true for very severe acute asphyxia, e.g., caused by total arrest of uterine blood flow, but also for mild to moderate asphyxia, caused by graded reduction in uterine blood flow. Furthermore, it shows that oxygen consumption is maintained as long as oxygen extraction can be increased (Wilkening and Meschia 1983). Only when oxygen extraction is maximal and when oxygen delivery continues to

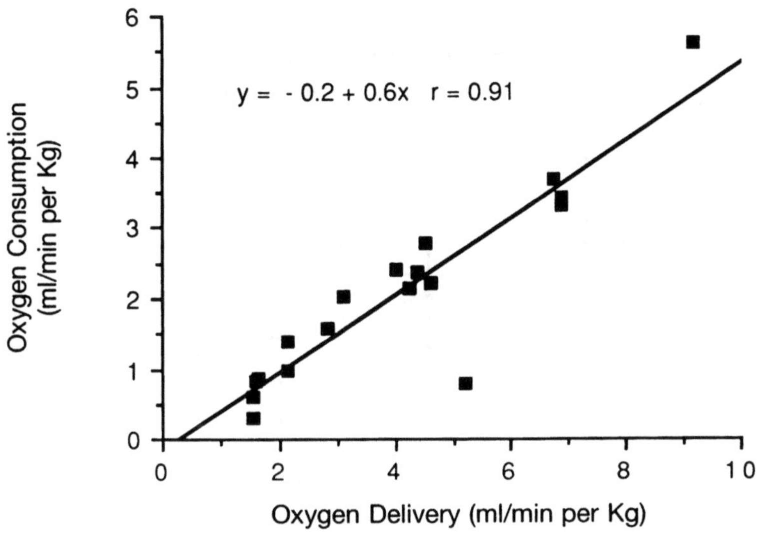

Fig. 22. Positive correlation between oxygen delivery to and oxygen consumption of the lower body segment in fetal sheep near term

fall does oxygen consumption begin to fall. From that point on, any further decrease in oxygen delivery is followed by a similar decrease in oxygen consumption. This study confirms our previous results obtained in vivo during arrest of uterine blood flow (Jensen et al. 1987c) and in vitro (Braems and Jensen 1990).

It seems important to notice that the circulatory and metabolic changes observed during graded reduction in uterine blood flow (to achieve a reduction in fetal oxygen delivery by 50%) differ from those observed during reduction in umbilical blood flow (Itskovitz et al. 1987). During reduction in umbilical blood flow that caused a similar reduction in fetal oxygen delivery, i.e., to 50% of control, blood flow to the carcass increased rather than decreased (Fig. 23b). These obvious differences in the distribution of blood flow to the fetal organs caused by uterine and by umbilical blood flow reduction may be related to carotid arterial oxygen content and pCO_2, which were different in the two studies. There was a fall in oxygen content and a steep rise of pCO_2 in the carotid artery during reduction in uterine blood flow, while there was only a small or no change in these variables during reduction in umbilical blood flow (Fig. 23a). This suggests that also carbon dioxide, which is known to enhance the response of peripheral arterial chemoreceptors to hypoxia (Dawes et al. 1969), may increase vasoconstrictive responses in the peripheral vasculature, which are in part mediated by the sympathetic nervous system. This view is supported by the close negative correlation between carotid arterial pCO_2 and blood flow to the lower body segment during graded reduction in uterine blood flow (Fig. 24).

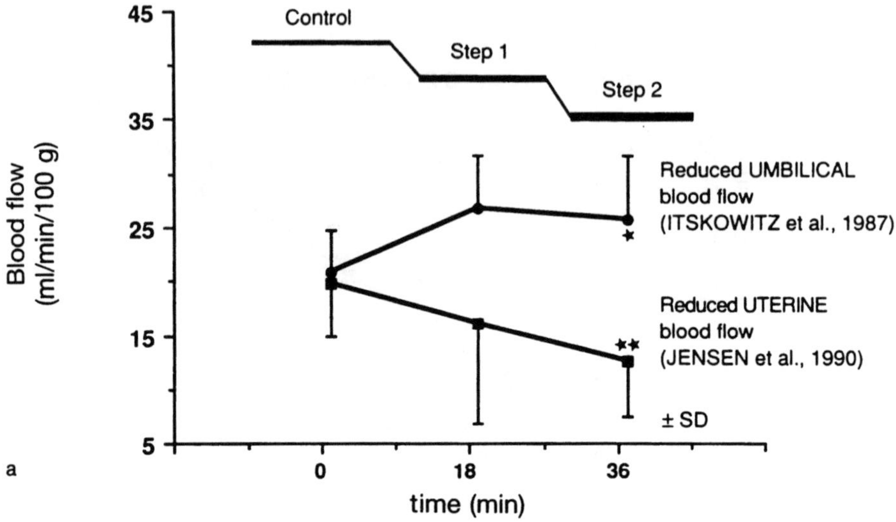

a

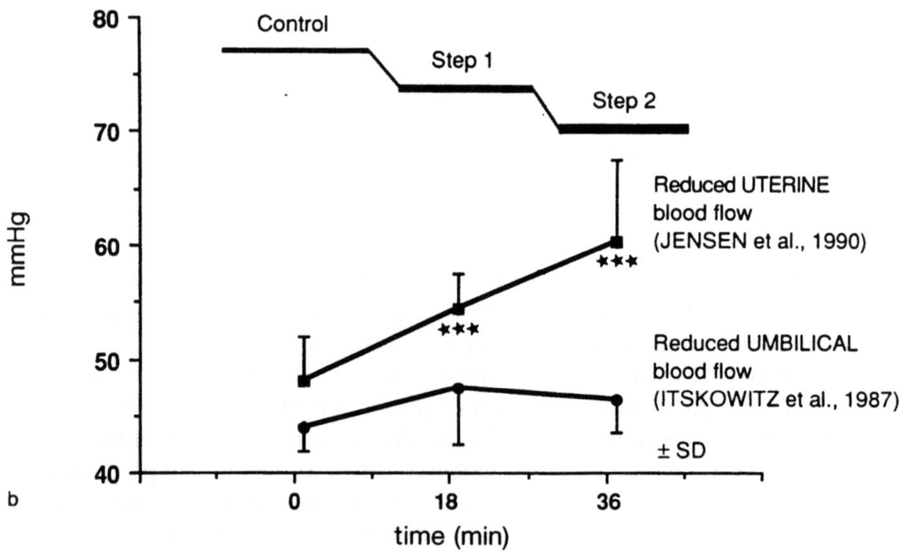

b

Fig. 23a, b. Changes in (a) carotid arterial carbon dioxide tensions (pCO_2) and in (b) blood flow to the lower carcass during of fetal sheep near term, reduction in total fetal oxygen delivery to 50% of control caused by a reduction of umbilical blood flow (Itskovitz et al. 1987) and by a reduction in uterine blood flow (Jensen et al. 1991). There are clear differences in both carbon dioxide tensions in the carotid artery and distribution of blood flow to the lower carcass, even though total oxygen delivery was reduced by the same amount. It is suggested that changes in carotid arterial pCO_2 may account in part for the circulatory differences observed, possibly by changing the sensitivity of peripheral arterial chemoreceptorsd. (After Itskovitz et al. 1987; and Jensen et al. 1991)

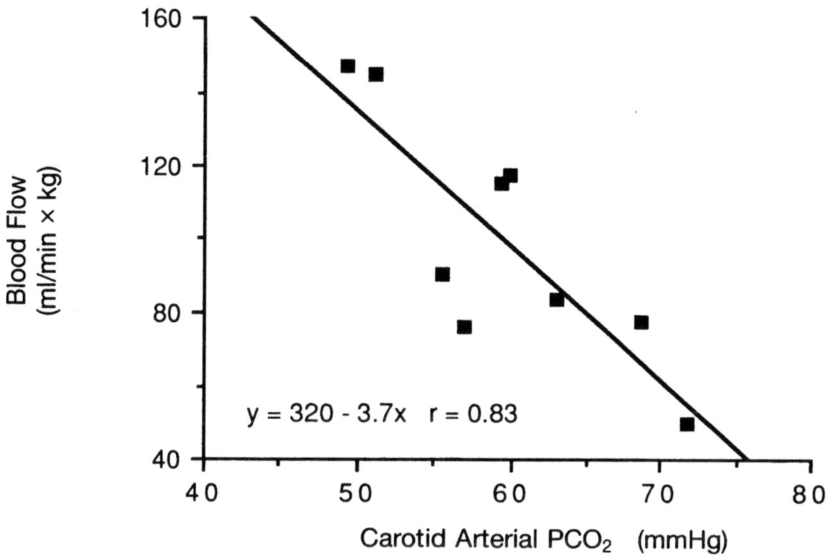

Fig. 24. Inverse correlation between blood flow to the lower body segment and carotid arterial pCO_2 in fetal sheep near term

Effects of Hypoxia on Oxygen Consumption in Fetal Cells in Monolayer Culture

Our previous results on the relation between oxygen delivery and oxygen consumption suggested that the fetus is able to adapt to situations in which oxygen is at short supply by reducing its oxygen consumption in peripheral organs. Furthermore, the close linear correlations between these variables supported the hypothesis that oxygen itself may be a regulator that controls oxygen consumption. However, there was no proof for this hypothesis. Rather the literature, which is largely based on studies in mitochondria preparations instead of whole cells, seemed to preclude such a possibility. We therefore designed an in vitro model to test this hypothesis, because we were unable to control important confounding factors in our in vivo model (Braems and Jensen 1990, 1991).

Methodologic details of the measurement in oxygen consumption in fetal monolayer culture are given elsewhere (Braems and Jensen 1990, 1991; this volume, p. 174). Briefly, specimens from the brain, myocardium, and skeletal muscle were obtained from guinea pig fetuses near term (term is at 68 days) under sterile conditions. These were rinsed with Hank's balanced salt solution (HBSS), cut into pieces, and treated with trypsin to dissociate fetal skeletal muscle cells. After centrifugation, the cells were resuspended in a medium containing basal medium (Eagle) with Hank's salts (BME), fetal calf serum, antibiotics, glutamine, and a buffer. Depending on the density of the cell population, cells were trypsinated and transferred to new flasks. Then the cells were transferred to Petri dishes, where they adhered and grew as a monolayer. The Petri dish was converted into a specially designed, air tight chamber, using a rubber ring and a metal plate with inlet and outlet manifolds, which allowed the salt solution to perfuse, or to bypass the cell culture. The air tight chamber was placed in a warm waterbath maintained at 37°C, using preincubated HBSS as perfusate. The pO_2 of the perfusate (flow rate 1 ml/min) was measured continuously during outflow, by a polarographic electrode (Transoxode, Dräger and Oxymonitor, Hellige, FRG). At the end of the experiment, the cells were dissociated from the Petri dish using trypsin. Both the number of dissociated cells and their vitality (usually more than 95%, trypan blue stain) were determined.

To determine the oxygen consumption of the cultured cells in the Petri dish, the difference in oxygen tensions across the Petri dish was measured under steady state conditions. Oxygen consumption per cell, per time, was calculated as follows: difference in pO_2 × flow rate × concentration of dissolved oxygen at a given measured pO_2 per number of cells.

To assess the change of cellular oxygen consumption during hypoxia, two stocks of perfusate were used. The control perfusate was equilibrated with air (~ 150 mmHg). Hypoxic perfusates were equilibrated with various gas mixtures containing reduced oxygen tensions, to achieve pO_2 levels between 10 and 140 mmHg without changing pH or pCO_2. The change of cellular oxygen consumption during hypoxia was expressed as a percent age of control.

During perfusion with control perfusate the absolute oxygen consumption of the brain cells (largely glia), myocardial cells, and skeletal muscle cells varied. However, in these different cell populations the relative changes (as percentage of control) of oxygen consumption during hypoxia were very similar. In all three populations there was a close positive correlation between oxygen delivered to and oxygen consumed by the cells (Braems and Jensen 1990, 1991; Braems et al. 1990; G. Braems, A. Peltzer, A. Jensen, unpublished work). A third degree polynomial could be fitted to each individual set of data points ($p \leq 0.001$).

These results, based on 20–30 experiments in each group, suggested that in the brain (glia), myocardium, and skeletal muscle of fetal guinea pigs near term oxygen is an important determinant of oxygen consumption. Furthermore, these results from our in vitro model which excluded any blood-borne or circulatory influences, confirmed our previous observations in vivo (Jensen et al. 1987c).

Thus, during fetal life on transition from normoxia to hypoxia the fetus is able to reduce oxygen consumption by decreasing oxygen delivery to peripheral organs, e.g., skeletal muscle. This – along with increasing blood flow to central organs – helps to maintain oxygen consumption in central organs by maintaining oxygen delivery to the brain and heart. Conversely, on transition from hypoxia to normoxia the increase in oxygen delivery is paralleled by an increase in cellular oxygen consumption and in metabolic drive in most fetal organs. This mechanism, which may be of particular importance on transition from fetal to postnatal life, when oxygen delivery and oxygen consumption rise (Dawes and Mott 1959), warrants optimal cell function at any given state of oxygenation.

Summary

In summary, we conclude that near term the fetal sympathetic nervous system plays an important part in preventing the adverse effects of asphyxia on the integrity of the brain, by causing both a circulatory and a metabolic centralization. Secondly, we conclude that the immaturity of the sympathetics before term may explain in part why premature babies are more susceptible to asphyxial brain damage than are mature babies. And finally, we conclude that central and peripheral fetal organs are able to adapt to hypoxia through a powerful cellular mechanism, by which oxygen consumption is reduced along with a reduction in oxygen delivery. During hypoxia and asphyxia, this mechanism acts in concert with a circulatory centralization which maintains oxygen delivery to, oxygen consumption of, and cell function in the brain and heart to ensure intact survival of the fetus (Jensen and Berger 1991).

Acknowledgements. The author gratefully acknowledges the invaluable help of Prof. W. Künzel and all the contributors to the various studies, which have been generously supported by the Deutsche Forschungsgemeinschaft.

The studies and researchers include: ultrasound screening of the brain, with V. Klingmüller, S. Sefkow, and S. Bender; the study of energy-rich phosphates, with R. Berger, J. Krieglstein, and J.-P. Steigelmann; blood flow studies, with W. Künzel, U. Lang, G. Braems, and M. Hohmann; cellular oxygen consumption, with G. Braems, K. Valentin, A. Peltzer, and I. Dussler; HPLC catecholamine analyses, with J. Jelinek, H. Klossek, and H.-J. Klönne; and technical assistance by S. Jelinek, B. Maid, R. Achenbach, D. Franke, and O. Adam.

References

Acheson GH, Dawes GS, Mott JC (1957) Oxygen consumption and the arterial oxygen saturation in foetal and new-born lambs. Physiol (Lond) 135: 623–642

Acker H, Lübbers DW (1977) The kinetics of local tissue pO_2 decrease after perfusion stop within the carotid body of the cat in vivo and in vitro. Pflügers Arch 369: 135–140

Ashwal S, Majcher JS, Vain N, Longo LD (1980) Patterns of fetal lamb regional cerebral blood flow during and after prolonged hypoxia. Pediatr Res 14: 1104–1110

Ashwal S, Majcher JS, Longo LD (1981) Patterns of fetal lamb regional cerebral blood flow during and after prolonged hypoxia: studies during the posthypoxic recovery period. Am J Obstet Gynecol 139 (4): 365–372

Barcroft J, Kennedy JA, Mason MF (1939) The direct determination of the oxygen consumption of the foetal sheep. Physiol (Lond) 95: 269–275

Berger R, Jensen A, Krieglstein J, Steigelmann J-P (1991) Effects of acute asphycia on brain energy metabolism in fetal guinea pigs near term I Dev Physiol, 16, 9–11

Berger R, Jensen A, Krieglstein J, Steigelmann J (1990) Zerebraler Sauerstoffmangel des Feten: Dynamische Veränderungen energiereicher Phosphate und Glykolyseintermediate bei akuter Asphyxie. 162th Meeting of the Mittelrheinische Gesellschaft für Gynäkologie und Geburtshilfe, Bad Dürkheim, FRG

Bernthal T, Schwind FJ (1945) A comparison in intestine and leg of the reflex vascular response to carotid-aortic chemoreceptor stimulation. Am J Physiol 143: 361–372

Braems G, Jensen A (1991) Hypoxia reduces oxygen consumption of fetal skeletal muscle cells in monolayer culture J Dev Physiol 16, 209–215

Braems G, Dussler I, Jensen A (1990) Oxygen availability determines oxygen consumption of fetal myocardial cells in monolayer culture. Meeting of the Society for Gynecologic Investigation, 21–24 March 1990, St Louis

Bristow J, Rudolph AM, Itskovitz J, Barnes R (1983) Hepatic oxygen and glucose metabolism in the fetal lamb. Response to hypoxia. J Clin Invest 71: 1047–1061

Campbell AGM, Dawes GS, Fishman AP, Hyman AI, James GB (1966) The oxygen consumption of the placenta and foetal membranes in the sheep. Physiol (Lond) 182: 439–464

Campbell AGM, Dawes GS, Fishman AP, Hyman AI (1967) Regional redistribution of blood flow in the mature fetal lamb. Circ Res 21: 229–235

Cohn HE, Sacks EJ, Heymann MA, Rudolph AM (1974) Cardiovascular responses to hypoxemia and acidemia in fetal lambs. Am J Obstet Gynecol 120 (6): 817–824

Creasy RK, Drost M, Green MV, Morris JA (1970) Determination of fetal, placental and neonatal blood volumes in the sheep. Circ Res 27 (4): 487–494

Dawes GS, Mott JC (1959) The increase in oxygen consumption of the lamb after birth. J Physiol (Lond) 146: 295–315

Dawes GS, Lewis BV, Milligan JE, Roach MR, Talner NS (1968) Vasomotor responses in the hind limbs of foetal and new-born lambs to asphyxia and aortic chemoreceptor stimulation. J Physiol (Lond) 195: 55–81

Dawes GS, Duncan SL, Lewis BV, Merlet CL, Owen-Thomas JB, Reeves JT (1969) Hypoxaemia and aortic chemoreceptor function in foetal lambs. J Physiol (Lond) 201: 105–116

De Burgh Daly M, Scott MJ (1963) The cardiovascular responses to stimulation of the carotid body chemoreceptors in the dog. J Physiol (Lond) 165: 179–197

Hanson MA, Jensen A (1989) Cardiovascular responses to acute asphyxia in carotid sinus denervated and intact fetal sheep near term. J Physiol (Lond) 417: 19P

Heymann MA, Payne BD, Hoffman JIE, Rudolph AM (1977) Blood flow measurements with radionuclide – labeled particles. Prog Cardiovasc Dis 20 (1): 55–79

Itskovitz J, La Gamma EF, Rudolph AM (1987) Effects of cord compression on fetal blood flow distribution and O_2 delivery. Am J Physiol 252: H100–H109

Iwamoto HS, Rudolph AM, Mirkin BL, Keil LC (1983) Circulatory and humoral responses of sympathectomized fetal sheep to hypoxemia. Am J Physiol 245: H767–H772

Jelinek J, Jensen A (1988) Distribution of catecholamines in central and peripheral organs of the fetal guinea pig during normoxaemia, hypoxaemia and asphyxia. In: Künzel W, Jensen A (eds) The endocrine control of the fetus – physiologic and pathophysiologic aspects. Springer, Berlin Heidelberg New York, pp 157–170

Jelinek J, Jensen A (1991) Catecholamine concentrations in plasma and organs of fetal guinea pigs during normoxaemia, hypoxaemia, and asphyxia. J Dev Physiol 15: 145–152

Jensen A (1989) Die Zentralisation des fetalen Kreislaufs. Thieme, Stuttgart

Jensen A, Berger R (1991) Fetal circulatory responses to oxygen lack. J Dev Physiol 16, 181–207

Jensen A, Hanson MA (1989) Brain blood flow and cardiovascular responses to acute asphyxia in carotid sinus denervated and intact fetal sheep near term. UPS Satellite Symposium, Fetal and Neonatal Physiology, p 120 (abstr no C9)

Jensen A, Künzel W (1980) The difference between fetal transcutaneous pO_2 and arterial pO_2 during labour. Gynecol Obstet Invest 11: 249–264

Jensen A, Lang U (1988) Dynamics of circulatory centralization and release of vasoactive hormones during acute asphyxia in intact and chemically sympathectomized fetal sheep. In: Künzel W, Jensen A (eds) The endocrine control of the fetus – physiologic and pathophysiologic aspects. Springer, Berlin Heidelberg New York, pp 135–149

Jensen A, Lang U (1989) Brain blood flow and the dynamics of circulatory centralization during acute asphyxia in intact and chemically sympathectomized fetal sheep. In: Krieglstein J (ed) Pharmacology of cerebral ischaemia 1988. Wissenschaftliche Verlagsgesellschaft, Stuttgart, pp 379–387

Jensen A, Schumacher R (1987) Intrakranielle Blutungen: Ursache, Diagnostik und Bedeutung. Gynäkologe 20: 52–64

Jensen A, Hohmann M, Künzel W (1987a) Redistribution of fetal circulation during repeated asphyxia in sheep: effects on skin blood flow, transcutaneous pO_2, and plasma catecholamines. J Dev Physiol 9: 41–55

Jensen A, Hohmann M, Künzel W (1987b) Dynamic changes in organ blood flow and oxygen consumption during acute asphyxia in fetal sheep. J Dev Physiol 9: 543–559

Jensen A, Künzel W, Kastendieck E (1987c) Fetal sympathetic activity, transcutaneous pO_2, and skin blood flow during repeated asphyxia in sheep. J Dev Physiol 9: 337–346

Jensen A, Hohmann M, Künzel W (1988) Die Änderung des fetalen und plazentaren Sauerstoffverbrauchs im Kreislaufschock: Sauerstoff ist ein Regulans. In: Dudenhausen JW, Saling E (eds) Perinatale Medizin, vol 12. Thieme, Stuttgart, pp 104–105

Jensen A, Klingmüller V, Sefkow S (1989a) Hirnblutungsrisiko bei Früh- und Reifgeborenen: Eine prospektive sonographische Reihenuntersuchung an 2781 Neugeborenen. In: Künzel W, Kirschbaum M (eds) Giessener Gynäkologische Fortbildung 1989. Springer, Berlin Heidelberg New York, pp 65–68

Jensen A, Klingmüller V, Künzel W, Sefkow S (1992) Das Hirnblutungsrisiko bei Früh- und Reifgeborenen. Geburtshilfe und Frauenheilkunde 52, 6–20

Jensen A, Lang U, Braems G (1989b) Dynamics of circulatory centralization during acute asphyxia in preterm and term fetal sheep. 36th Annual Meeting of the Society for Gynecologic Investigation, San Diego, p 153 (Scientific abstracts, no 145)

Jensen A, Roman C, Rudolph AM (1991) Effects of reducing uterine blood flow on fetal blood flow distribution and oxygen delivery. J Dev Physiol 15, 309–323

Killip T III (1963) Sinus nerve stimulation in the Chloralose anesthetized cat: effect on blood pressure, heart rate, muscle blood flow and vascular resistance. Acta Physiol Scand 57: 437–445

Myers RE, Yamaguchi M (1976) Effects of serum glucose concentration on brain response to circulatory arrest. J Neuropathol Exp Neurol 35: 301

Pape KE, Wigglesworth JS (1979) Haemorrhage, ischaemia and the perinatal brain. Lavenham Press, Lavenham, UK

Papile LA, Burstein J, Burstein R, Koffler H (1978) Incidence and evolution of subependymal and intraventricular hemorrhage: a study of infants with birth weights less than 1.500 gm. J Pediatr 92: 529

Paulick R, Kastendieck E, Wernze H (1985) Catecholamines in arterial and venous umbilical blood: placental extraction, correlation with fetal hypoxia, and transcutaneous partial oxygen tension. J Perinat Med 13: 31–42

Peeters LLH, Sheldon RE, Jones MD, Makowski EL, Meschia G (1979) Blood flow to fetal organs as a function of arterial oxygen content. Am J Obstet Gynecol 135: 637–646

Rudolph AM, Heymann MA (1967) The circulation of the fetus in utero: methods for studying distribution of blood flow, cardiac output and organ blood flow. Circ Res 21: 163–184

Siesjö BK (1981) Cell damage in brain: a speculative synthesis. J Cereb Blood Flow Metab 1: 155–185

Wilkening RB, Meschia G (1983) Fetal oxygen uptake, oxygenation, and acid-base balance as a function of uterine blood flow. Am J Physiol 244: H749–H755

Wilson DF, Owen CS, Erecinska M (1979) Quantitative dependence of mitochondrial oxidative phosphorylation on oxygen concentration: a mathematical model. Arch Biochem Biophys 195 (2): 494–504

The Response and Vulnerability of Cerebral Blood Vessels to Fetal Hypoxemia

M. KIRSCHBAUM, A. KRIETE, R. H. BÖDEKER, and W. KÜNZEL

Introduction

During the early part of this century, cranial trauma and asphyxia during birth were the most frequent phenomena in perinatal pathology. Atraumatic bleeding into the ventricles was regarded as an uncommon phenomenon in term pregnancies, but as early as 1919 Ylppö characterized bleedings into the ventricles as a typical sign of hemorrhage in the preterm infant. Bleeding into the subependymal matrix near the foramen of Monro (Hambleton and Wiggelsworth 1976) was often found to be accompanied by a rupture and bleeding into the lateral ventricle which then spread throughout the ventricular system (Hemsath 1934). With the general decline in perinatal mortality in the past decades, the relative importance of this type of preterm lesion has increased (Lee et al. 1980).

An additional cause of brain damage is the reduced oxygen supply to the fetus induced by chronic placental insufficiency with fetal growth retardation (Naeye 1986; Niswanders 1983), maternal alcoholism, drug addiction, and maternal and placentofetal infections.

Despite gross brain pathology it must be recognized that in the injured brain there is a measurable decrease in the number of nerve cells, dendrites, and synapses (Larroche 1984). These effects, together with a delay of and anomalies in myelination (Leviton et al. 1983), might be the morphological correlate of a great number of minor and major brain dysfunctions.

The pathway of the development of fetal brain injury is shown in Fig. 1. In the acute asphyxiated fetus, blood pressure rises, and blood is redistributed to the heart and the brain. Parallel to the redistribution of blood flow, the O_2 supply to the brain is impaired, and the question arises what impact this has on brain cells and the vasculature of the brain. Studies of acute hypoxic episodes in a chronic sheep preparation have shown that the choroid plexus exhibits a tremendous vasoconstriction (Jensen et al. 1987). With the progression of lactate accumulation, moreover, vascular autoregulation is impaired (Milligan and Bryan, 1979). In the process of brain damage, the impairment of the microcirculation (Lou et al. 1979) plays an important role (Schulte 1988). The aim of this chapter is to describe the reaction and the change of permeability of the brain vasculature in response to asphyxia and as early sign of vascular lesions.

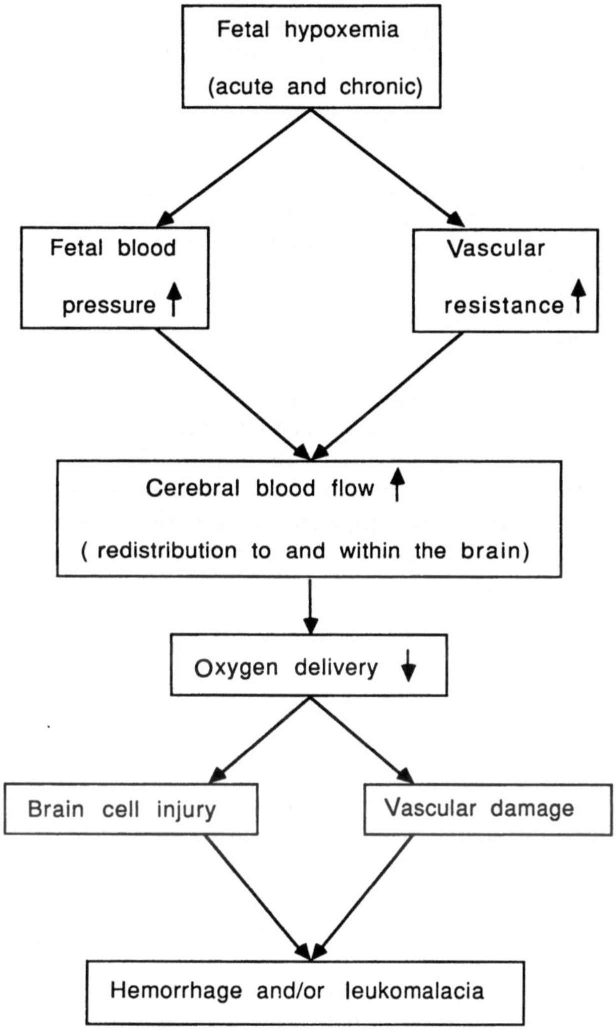

Fig. 1. The pathways of development of fetal brain injury (according to Volpe 1981)

The Effects of Fetal Hypoxia on Blood Vessel Diameter in Different Areas of the Fetal Brain

Experimental Design

Five guinea pig fetuses near term were exposed to intrauterine asphyxia by dripping oxytocine-solution on to the myometrium. After asphyxia had been induced, the ten brain hemispheres were fixed in formalin, and sections were stained according to the method of Pickworth (1934) to permit visualization of the blood vessels. Ten well-defined areas of the cerebrum were selected:

1. The cerebral cortex – specifically, the pars parietalis and
2. The gyrus cinguli.
3. The hippocampus.
4. The caudate nucleus – specifically, the head.
5. The thalamus – specifically, the medial and lateral nucleus, and
6. The ventral posterior and lateral posterior nucleus.
7. The hypothalamus.
8. The mesencephalon – specifically, the lamina tecti.
9. The cerebellum.
10. The pons.

The vessel cross sections were counted in ten fields per area in an image-analyzing system (KONTRON IBAS 2000, Kontron Elektronik GmbH, Eching, FRG) to determine the diameter of all vessels. From a section stained with hematoxilin and eosin (H&E) (Fig. 2), the brain areas were identified. The adjacent brain section, stained by Pickworth's method, is shown in Fig. 3. The blood vessels are shown at higher magnification in Figs. 4 and 5. The sections of the vessels identified were distinguished by the computer, and the minimal diameter drawn through the center of the vessel cross section was measured automatically.

Results

The results are based on the evaluation of about 300 000 blood vessels. The distribution of blood vessels of different sizes in the gyrus cinguli of the cortex of two different animals is shown in a histogram (Fig. 6). In the asphyxiated brain, the number of visible vessels is higher than in the control. There was also a change in the mean diameter of vessels in the asphyxiated fetus from 8 μm to 8.5 μm.

The number of sectioned vessels, the geometric mean, and the median of these vessels was used to describe the impact of fetal hypoxia. Blood gas analysis was performed from mixed blood obtained through heart punction after a period of asphyxia immediately before the brain was fixed. The parameters of the blood gas analysis, i.e., pH, base excess (BE), pO_2, and pCO_2, were related to the morphometrical parameters, which were calculated by the tissue analyzing system. This chapter shows the relationship between BE and the above mentioned morphometrical parameters.

In the gyrus cinguli, for example, the total number of blood vessels was related to the BE. With falling BE, the number of visible vessels rose (Fig. 7). In many brain areas, e.g., pons and mesencephalon, the degree of correlation was weak. However, in almost all brain areas there was a trend towards spare vessels opening with increasing asphyxia. A strong dependence exists between BE and the number of visible vessels in the gyrus cinguli of the cortex and the caudate nucleus (Fig. 8).

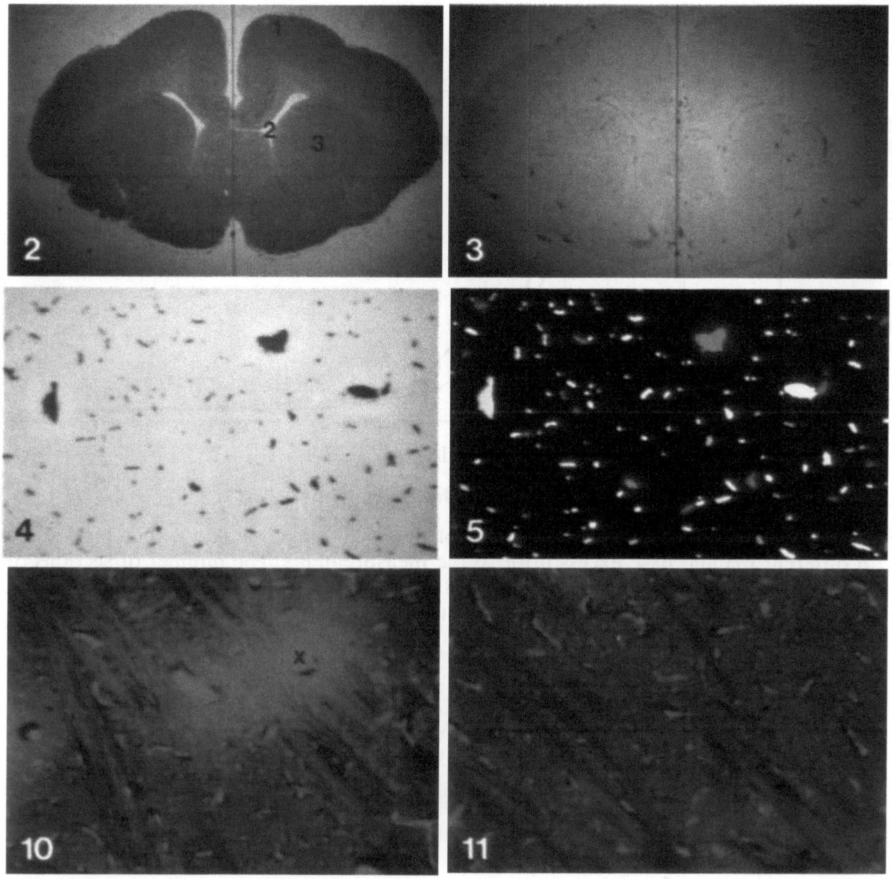

Fig. 2. Cross section of the brain in the guinea pig fetus. At the top, the cortex is shown (*1*). The lateral ventricle (*2*) and the caudate nucleus (*3*) can be seen on each hemisphere, hematoxilin and eosin, × 12

Fig. 3. Cross section of the brain in the guinea pig fetus, as in Fig. 1, stained after the method of Pickworth. Only blood vessels (i.e. erythrocytes) are stained. They provide a good contrast to the unstained brain parenchyma. The blood vessels can easily be distinguished by the image-analyzing system, × 12

Fig. 4. Binary image of vessel cross sections of the fetal brain in the guinea pig, created by the analyzing system and as seen on the TV monitor. This image is generated in the computer by filming the histological section with the TV camera. The image is then digitalized with the image-analyzing system IBAS 2000 after various steps of image improvement such as shading correction, contrast enhancement, and segmentation by thresholding, × 40

Fig. 5. Binary image, as in Fig. 3 but after pseudocolors have been attributed to individual vessels. The pseudocolors ensure correct identification of the vessels by the analyzing system, × 40

Fig. 10. Fluorescent microscopic image showing high amont of extravasal Evans blue around the cerebral blood vessels (*x*) (guinea pig fetus 86-01; pH 6.91; base excess, −13.6 mEq/l; exciting illumination wavelength, 530–560 nm; barrier filter 580 nm, Leitz Ploemopak N2Y; × 40)

Fig. 11. Brain with low amount of extravasal Evans blue around the cerebral blood vessels (guinea pig fetus 86-03; pH 7.30; base excess, +4.0 mEq/l; × 40)

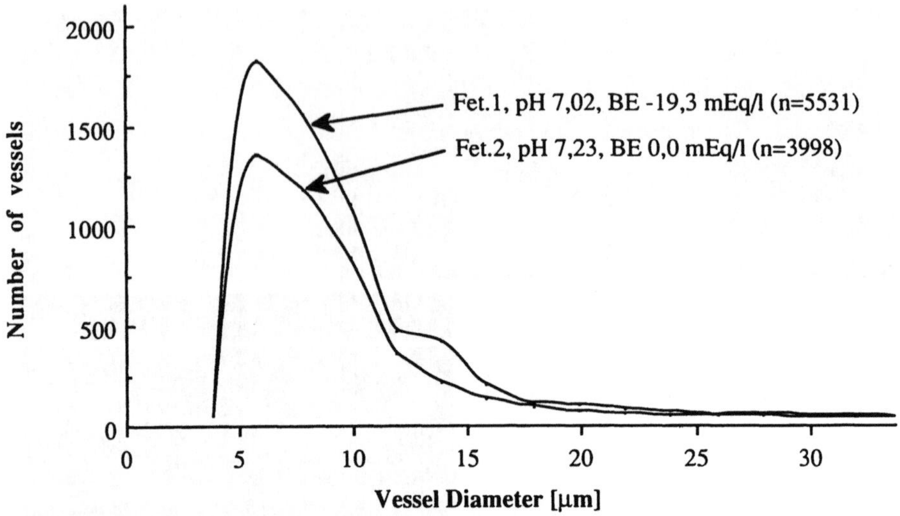

Fig. 6. The distribution of vessel diameters of the gyrus cinguli in two guinea pig fetuses with different pH and base excess prior to the measurements. The number of sections of blood vessels counted and measured was 3998 in the nonasphyxiated fetus and 5531 in the asphyxiated fetus. In the acidotic fetus, the number of vessels increased in parallel with a rise in the vessel diameter

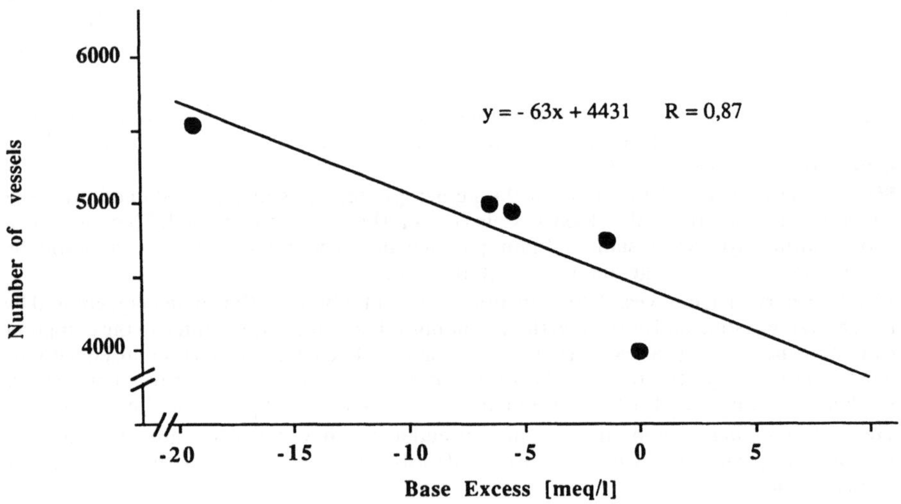

Fig. 7. The relationship between base excess in fetal blood and the number of vessels in the gyrus cinguli of five fetuses. There is a close correlation between both parameters showing that, with the fall in fetal base excess, the number of visible brain vessels increases, i.e., nonperfused and, therefore, invisible spare vessels are opened to improve cerebral circulation

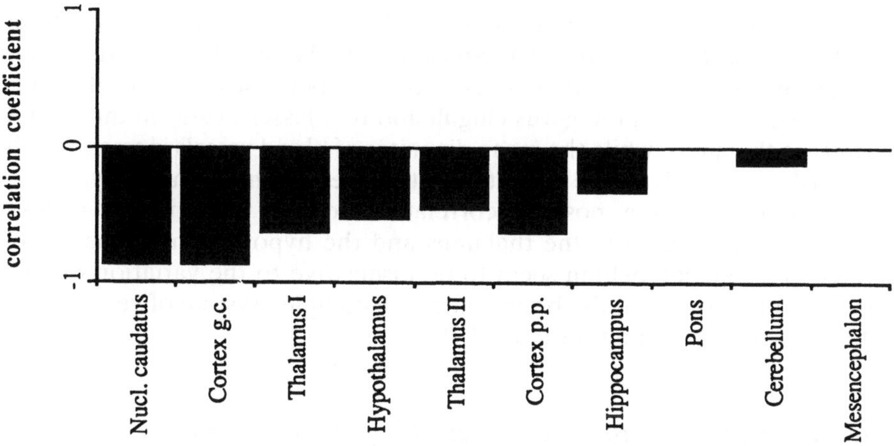

Fig. 8. The correlation coefficient for the relation between number of vessels and base excess in the fetal blood for ten different brain areas. A significant correlation between vessel number and base excess could be established for the following brain areas: caudate nucleus, cortex, thalamus, and hypothalamus, but not for the hippocampus, pons, cerebellum, and mesencephalon. This demonstrates that the various brain areas respond differently following fall of base excess

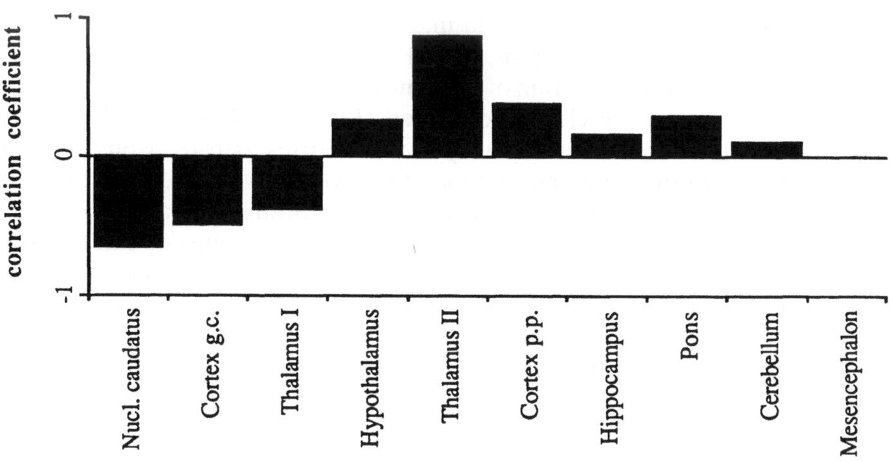

Fig. 9. The correlation coefficient for the relation between geometric mean and base excess in fetal blood. The distribution of the correlation coefficient related to the geometric mean differs obviously from that of the absolute number of cerebral blood vessels. In the caudate nucleus, there is a strong tendency towards dilatation of vessels (increasing geometric mean of vessel diameters) with falling base excess, whereas in the occipital part of thalamus (thalamus II) there is vasoconstriction at the same time

A different correlation exists between the mean diameter of the blood vessels and the BE in various brain areas (Fig. 9). With falling BE, the mean vascular diameter rises in the vessels of the caudate nucleus. This phenomenon is also obvious in the gyrus cinguli and to a lesser extent in the rostral part of the thalamus, while the mean diameter of blood vessels increases with falling BE, especially in the occipital part of the thalamus. In the hypothalamus, there is also a positive correlation to BE, i.e., if BE is low, vasoconstriction occurs in the thalamus and the hypothalamus. The vessel diameters in the cerebellum seem to be insensitive to the variation of BE. These results reveal that the brain contains a complex system of vessels with differing sensitivity to alterations of BE.

The Permeability of the Blood Vessels of the Fetal Brain in Response to Fetal Deterioration

Experimental Design

A more serious phenomenon following asphyxia is the vascular damage in the fetal brain. It is supposed that the damage of the vessels occurs gradually, leads finally to the breakdown of the blood-brain barrier (Ito et al. 1976), and ends in cerebral hemorrhage. It may be speculated that the cascade is reversible to a certain point, but if it continues irreversible damage takes place. We wanted to investigate whether there was any evidence for the opening of tight junctions in the endothelium of the cerebral vessels, for the penetration of plasmatic proteins into the cerebral parenchyma, and for the development of intracerebral edema under the conditions of asphyxia.

To quantify the extravasation, the Evans blue technique was used. Histological sections of the brain of a normoxic fetal brain and the brain of an asphyxiated fetus can be clearly identified (Figs. 10, 11, page 112). Evans blue was administered intravenously to 28 fetuses before asphyxia and visualized in a fluorescent microscope. The dye shows a red fluorescence when illuminated by green light. The asphyxiated brain emits a strong Evans blue fluorescence in comparison to the control. The amount of extravasally located Evans blue was measured and quantified by the time needed for a certain number of photons to reach a photo sensor. Short times are associated with high intensity and long times with low intensity of Evans blue in the brain area. The data were expressed as a permeability index (PI), with $PI = 10 (\log 1/t + 2)$.

Results

With increasing acidosis in the fetal brain, the permeability of blood vessels for Evans blue rises (Fig. 12). This phenomenon can be observed in all cerebral regions. Evans blue binds immediately to albumin if given intrave-

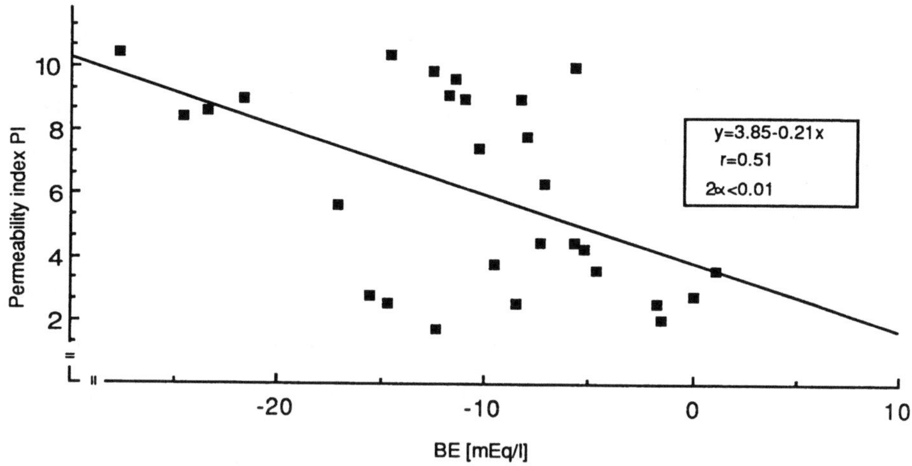

Fig. 12. Evans blue permeability index (PI) correlated to base excess in fetal blood. There is a good correlation between the fall in base excess and the increase in Evans blue extravasates, e.g., with falling base excess, the permeability of cerebral blood vessels to Evans blue rises

nously. It acts therefore as a tracer for albumin. The opening of the blood-brain barrier to Evans blue is a sign of the penetration of albumin from the intravasal into the intracerebral compartment *and* of cerebral edema, with all its possible consequences, namely, that the distance becomes longer for oxygen transfer to astrocytes and ganglion cells, which may lead to nerve cell death, and to a further increase in vascular leakage, also for blood cells. Although the adult brain can compensate for brain edema without detrimental consequences, the fetal brain insult may end in a cerebral dysfunction.

The results presented here indicate that it is important to prevent perinatal hypoxia and asphyxia, even if not all cases of brain asphyxiation lead to measurable nerve cell death or cerebral bleeding.

References

Hambleton G, Wigglesworth JS (1976) Origin of intraventricular hemorrhage in the preterm infant. Arch Dis Child 51:651–659

Hemsath FH (1934) Birth injury of the occipital bone with a report of 32 cases. Am J Obstet Gynecol 22:194–203

Ito U, Go KG, Walker JT, Spatz M, Klatzo I (1976) Experimental cerebral ischaemia in Mongolian gerbils. III. Behaviour of the blood-brain barrier. Acta Neuropathol (Berl) 34:1–6

Jensen A, Hohmann M, Künzel W (1987) Dynamic changes in organ blood flow and oxygen consumption during acute asphyxia in fetal sheep. J Dev Physiol 9:543–559

Larroche JC (1984) Perinatal brain damage. In: Adams JH, Corsellis JAN, Duchen LW (eds) Greenfield's neuropathology. Arnold, London

Lee KS, Paneth N, Gartner L (1980) Neonate mortality: an analysis of the recent decline in the United States. Am J Public Health

Leviton A, Gilles FH, Dooling EC (1983) The epidemiology of delayed myelination. In: Giles FH et al. (eds) Developing human brain, growth and epidemiologic neuropathology. Wright, Boston

Lou HC, Lassen NA, Tweed WA, Johnson G, Jones M, Palahniuk RJ (1979) Pressure passive cerebral blood flow and break down of the blood-brain barrier in experimental fetal asphyxia. Acta Paediatr Scand 68:35

Milligan DW, Bryan MH (1979) Failure of autoregulation of the cerebral circulation in the sick newborn infant. Paediatr Res 13:527

Naeye RL (1986) When and how does antenatal brain damage occur? In: Iffy L (ed) 2nd perinatal practice and malpractice Symposium. Healthmark Community, New York, p 125

Niswander KR (1983) Asphyxia in the fetus and cerebral palsy. Yearbook, Chicago, pp 107–125

Pickworth FA (1934) A new method of study of the brain capillaries and its application to the regional localisation of mental disorder. J Anat 69:62–71

Schulte FJ (1988) Pathophysiological mechanisms leading to permanent brain damage in surviving children. In: Kubli F et al. (eds) Perinatal events and brain damage in surviving children. Springer, Berlin Heidelberg New York

Volpe JJ (1981) Neurology of the newborn. Saunders, Philadelphia

Ylppö A (1919) Pathologisch-anatomische Studien bei Frühgeburten. Z Kinderheilkd 20:212–431

Breathing Movements in the Human Fetus in Normoxia and Hypoxia

J. G. Nijhuis

Introduction

It was back in 1888 and 1905 that Ahlfeld published his observations of abdominal wall movements in pregnant women, which he interpreted as being caused by breathing movements carried out by the fetus. He said: "Die Stöße treten im Durchschnitt alle drei bis vier Sekunden auf bei Schwankungen der Periodizität zwischen 1 und 32 Sekunden. Sie kennzeichnen sich als schnell auf- und absteigende Linien, also als ganz kurze Stöße" (The movements appear on average every 3–4 s, although fluctuations in intervals can occur where the movements might appear anywhere between 1 and 32 s. They are characterized as quick rising and falling contours or very short thrusts) (Ahlfeld 1888, p. 205). He also noticed that these movements are not continuously present: "Gerade in den letzten Wochen habe ich viele Schwangeren vergeblich untersucht, ohne einmal den Zeitpunkt zu treffen, wo dauernd derartige Stöße zu beobachten gewesen wären" (Especially in the last weeks of pregnancy, I have examined many women without ever once coming across the circumstance where these kinds of movements were constantly to be observed) (Ahlfeld 1888, p. 204). After a period of nearly unanimous neglect of Ahlfeld's observations, the issue was re-introduced by Dawes and co-workers in 1970. They described breathing movements in the fetal lamb; and 1 year later Boddy and Robinson (1971) reported on breathing movements in the human fetus, detected by A-mode ultrasound. Other techniques that have been used to detect human fetal breathing movements include external tocodynamometers (Timor-Tritsch et al. 1979), continuous wave Doppler ultrasound and real-time B-mode ultrasound (see Cousin 1980). Using this last technique, breathing movements can be observed from 11 weeks of gestation onward (De Vries et al. 1982).

In this chapter we summarize briefly the recent data on breathing movements with respect to their incidence under normal conditions, in hypoxia and in hyperoxia.

Incidence of Breathing Movements

The incidence of breathing movements reported in several studies varies greatly. One of the explanations is of course the definition of apnea that is

Table 1. Breathing incidence in the low-risk fetus

Menstrual weeks of gestation	Incidence[a] (%)	Apnea[b] (min)
24–26[c]	13.7	12
26–28[c]	14.2	14
30–31[d]	31.3	68
34–35[e]	31.8	105
38–39[d]	31.4	122

[a] Mean incidence over 24 h.
[b] Longest period of apnea observed. Apnea is defined as a period of at least 6 s, while periods with body movements are also excluded.
[c] Natale et al. 1988.
[d] Patrick et al. 1980b.
[e] Patrick et al. 1978.

used and the way recordings are analyzed, e.g., by epoch analysis or using a moving window technique. Patrick et al. (1980a) showed that apnea could well be defined by a breath-to-breath interval of 6 s or more. Data on incidence of fetal breathing movements and the longest periods without breathing movements are summarized in Table 1.

Especially in the last weeks of pregnancy breathing movements can cease over a long period of up to 120 min. Furthermore, the incidence of breathing movements increases after glucose intake (Natale et al. 1978) and 2–3 h after meals from 31 weeks of gestation (Patrick et al. 1980b). At 24–28 weeks this effect is also present after glucose intake (Nijhuis et al. 1986), but not after meals (Natale et al. 1988). Finally, there is a diurnal rhythm: from 31 weeks onward, the incidence of breathing movements increases between 04.00 and 07.00 h, while the mother is asleep. At 24–26 weeks, a peak occurrence is present between 23.00 and 02.00 hours, and at 26–28 weeks between 02.00 and 0.500 h (Patrick et al. 1980b; Natale et al. 1988).

Breathing and Behavioral States

In 1982 Nijhuis et al. introduced the concept of fetal behavioral states. It became clear that many physiological phenomena were state-dependent. Van Vliet et al. (1985a) showed that the incidence in state 2F (similar to rapid eye movement (REM) sleep, active sleep) was significantly greater than during state 1F (similar to non-REM (NREM) sleep, quiet sleep).

In the neonate, breathing is used as a state criterion. Breathing is very regular in state 1 (quiet sleep) and very irregular in state 2 (active sleep) (see Prechtl 1974).

In 1980 Timor-Tritsch et al. showed a similar relation for breathing movements and active and quiet periods in the human fetus. Nijhuis et al.

(1983), using the state concept, confirmed these results: fetal breathing – if present – is much more regular during state 1F than during state 2F.

Breathing and Hypoxia

Using the example of a woman with sickle cell anemia, in crisis, Manning and Platt (1979) gave evidence that human fetal breathing movements cease during acute hypoxic events. This phenomenon is well-known in animal experimental work (Boddy et al. 1974). In normal pregnancy it is difficult to measure a degree of hypoxia, and it is certainly unethical to induce hypoxia.

It is often assumed that intrauterine growth retardation (IUGR) may develop as a result of chronic hypoxia. Bekedam and co-workers studied IUGR fetuses within 24 h before cesarean sections. Fetuses with IUGR and late decelerations were indeed significantly more hypoxemic at birth than those in the group without decelerations (Bekedam et al. 1987). In IUGR fetuses, breathing incidence also decreased significantly after late deceleration (Bekedam and Visser 1985).

The fact that a late heart rate deceleration probably reflects an hypoxic event is another strong argument that in the human fetus breathing ceases during hypoxia. This effect may be more pronounced if IUGR or another situation of chronically mild hypoxia is already present, and if severe hypoxia is then superimposed.

Several research groups studied breathing in both normal and IUGR fetuses (Ritchie and Lakhani 1980; Fornan and Ritchie 1983; Devoe et al. 1984; Van Vliet et al. 1985a; Ruedrich et al. 1989). None of them found a significant difference between the two groups, but they all found a nonsignificant difference, namely, the incidence of breathing was decreased in all IUGR groups. This may indicate that there is a mild decrease of breathing movements, although nonsignificant, in chronically hypoxic fetuses.

Breathing and Hyperoxia

If one assumes that IUGR represents a hypoxic state and that fetuses breathe less during hypoxia, then one might hypothesize that the incidence of breathing would increase more during maternal hyperoxia in IUGR than in normal fetuses.

In low-risk fetuses the effect of extra oxygen has been tested by several groups, and an increase of breathing incidence was never noted. Three of these groups followed the same protocol in IUGR fetuses, where significant increases of breathing incidence were found (see Table 2). These findings support the idea that in the human growth-retarded fetus, breathing movements are mildly depressed during chronic hypoxia.

Table 2. Fetal breathing and the effect of maternal hyperoxia: summary of studies

Authors		Low-risk	IUGR
Ritchie and Lakhani	1980	No effect	Increase
Dornan and Ritchie	1983	No effect	Increase
Devoe et al.	1984	No effect	–
Van Vliet et al.	1985b	No effect	–
Ruedrich et al.	1989	No effect	Increase

Conclusions

The presence of breathing movements is in general a reliable indicator of a good fetal condition, but the absence of breathing movements does not necessarily have many implications. Breathing movements should always be studied under carefully standardized conditions to avoid influences of gestational age, meals, and diurnal rhythms.

In normal and IUGR fetuses there is no significant difference in the incidence of breathing, although most studies find a somewhat lower incidence in IUGR fetuses. If IUGR is a chronic hypoxic condition, then we must assume that the breathing incidence in chronic hypoxia is more or less similar to breathing in normoxia, or perhaps only mildly reduced. In acute hypoxia however, breathing movements probably cease in the human fetus, as is the case with the fetal lamb.

Finally, maternal hyperoxia may stimulate fetal breathing in utero. This is certainly the case in IUGR fetuses. Whether this observation could lead to the development of a useful clinical tool or test, however, is yet to be determined.

References

Ahlfeld F (1888) Über intrauterine Atembewegungen des Kindes. Verh Dtsch Ges Gynäkol 2:203–210

Ahlfeld F (1905) Die intrauterine Tätigkeit der Thorax- und Zwerchfellmuskulatur. Intrauterine Atmung. Monatsschr Geburtshilfe Gynäkol 21:143–163

Bekedam DJ, Visser GHA (1985) Effects of hypoxemic events on breathing, body movements, and heart rate variation: a study in growth-retarded human fetuses. Am J Obstet Gynecol 153:52–56

Bekedam DJ, Visser GHA, Mulder EJH, Poelmann-Weesjes G (1987) Heart rate variation and movement incidence in growth-retarded fetuses: the significance of antenatal late heart rate decelerations. Am J Obstet Gynecol 157:126–133

Boddy K, Robinson JS (1971) External method for detection of fetal breathing in utero. Lancet II:1231–1233

Boddy K, Dawes GS, Fisher RL, Robinson JS (1974) Foetal respiratory movements, electrocortical and cardiovascular responses to hypoxaemia in sheep. J Physiol (Lond) 243:599–618

Cousin AJ (1980) Advanced methods for measurement of fetal breathing movements in the human fetus. Semin Perinatol 4:261–268

Dawes GS, Fox HE, Leduc BM, Liggins GC, Richards RT (1970) Respiratory movements and paradoxal sleep in the foetal lamb. J Physiol (Lond) 210:47–48

De Vries JIP, Visser GHA, Prechtl HFR (1982) The emergence of fetal behaviour. I. Qualitative aspects. Early Hum Dev 7:301–322

Devoe LD, Abduljabbar H, Carmichael L, Probert C, Patrick J (1984) The effects of maternal hyperoxia on fetal breathing movements in third-trimester pregnancies. Am J Obstet Gynecol 148:790–794

Dornan JC, Ritchie JWK (1983) Fetal breathing movements and maternal hyperoxia in the growth retarded fetus. Br J Obstet Gynaecol 90:210–213

Manning FA, Platt LD (1979) Maternal hypoxemia and fetal breathing movements. Obstet Gynecol 53:758–760

Natale R, Patrick J, Richardson B (1978) Effects of human maternal venous plasma glucose concentrations on fetal breathing movements. Am J Obstet Gynecol 132:36

Natale R, Nasello-Paterson C, Connors G (1988) Patterns of fetal breathing activity in the human fetus at 24 to 28 weeks of gestation. Am J Obstet Gynecol 158:317–321

Nijhuis JG, Prechtl HFR, Martin CB Jr, Bots RSGM (1982) Are there behavioural states in the human fetus? Early Hum Dev 6:177–195

Nijhuis JG, Martin CB Jr, Gommers S, Bouws P, Bots RSGM, Jongsma HW (1983) The rhythmicity of fetal breathing varies with behavioural state in the human fetus. Early Hum Dev 9:1–7

Nijhuis JG, Jongsma HW, Crijns IJMJ, de Valk IMGM, van der Velden WHJ (1986) Effects of maternal glucose ingestion on human fetal breathing movements at weeks 24 and 28 of gestation. Early Hum Dev 13:183–188

Patrick J, Natale R, Richardson B (1978) Patterns of human fetal breathing activity at 34 to 35 weeks' gestational age. Am J Obstet Gynecol 132:507–513

Patrick J, Campbell K, Carmichael L, Natale R, Richardson B (1980a) A definition of human fetal apnea and the distribution of fetal apneic intervals during the last ten weeks of pregnancy. Am J Obstet Gynecol 136:471–477

Patrick J, Campbell K, Carmichael L, Natale R, Richardson B (1980b) Patterns of human fetal breathing during the las 10 weeks of pregnancy. Obstet Gynecol 56:24–30

Prechtl HFR (1974) The behavioural states of the newborn infant (a review). Brain Res 76:185–212

Ritchie JWK, Lakhani K (1980) Fetal breathing movements and maternal hyperoxia. Br J Obstet Gynaecol 87:1084–1086

Ruedrich DA, Devoe LD, Searle N (1989) Effects of maternal hyperoxia on the biophysical assessment of fetused with suspected intrauterine growth retardation. Am J Obstet Gynecol 161:188–192

Timor-Tritsch IE, Dierker LJ, Hertz RH, Zador T, Rosen MG (1979) Human fetal respiratory movements: a technique for non-invasive monitoring with the use of a tocodynamometer. Biol Neonate 36:18–24

Timor-Tritsch IE, Dierker LJ, Hertz RH, Chik L, Rosen MG (1980) Regular and irregular human fetal respiratory movements. Early Hum Dev 4:315–324

van Vliet MAT, Martin CB Jr, Nijhuis JG, Prechtl HFR (1985a) The relationship between breathing movements and fetal activity and behavioural states in normal and growth retarded fetuses. Am J Obstet Gynecol 153:582–588

van Vliet MAT, Creemers HMH, van Heyst AFJ, Martin CB Jr (1985b) The effect of maternal hyperoxia on breathing movements, body movements and rest-activity cycles of low-risk human fetuses. Eur J Obstet Gynecol Reprod Biol 20:221–228

Fetal Hypoxemia as a Signal for Parturition

J. R. G. Challis, A. D. Bocking, B. S. Richardson,
A. N. Brooks, A. Sue-Tang, K. Akagi, and R. A. Jacobs

Introduction

Alterations in oxygenation are known to influence directly prostaglandin (PG) production by tissues such as fetal and neonatal lung cells, and by the fetal brain. However, at the present time there is no systematic study of the effects of altered oxygen tensions on PG output by the placenta or fetal membranes, the major sites of eicosanoid generation in the rocesses leading to birth. In comparison, there is considerable evidence that fetal hypoxemia activates components of the fetal hypothalamic-pituitary-adrenal-placental (HPAP) axis, including placental peptide output, and thereby may contribute indirectly to the stimulus to elevated PG generation and myometrial contractility. In this paper we shall consider situations of reduction in oxygenation that may influence HPAP activity. These include alterations in basal partial pressure of oxygen (PaO_2) with gestational age, or short-term fluctuations in PaO_2 resulting from episodic contractions of the myometrium throughout pregnancy. Second, there may be short-term acute reductions in fetal PaO_2, reproduced experimentally in various models of acute fetal hypoxemia. Third, there may be a chronic longer-term impairment of fetal oxygenation, contributing to intrauterine growth retardation and studied experimentally in models of placental embolism, or chronic reductions in uteroplacental blood flow.

Basal and Short-Term Fluctuations in PaO_2

In samples of fetal umbilical venous and arterial blood obtained by cordocentesis, there is a progressive decrease in PO_2 and increase in carbon dioxide pressure (PCO_2) in appropriate-for-gestational age fetuses with progression of gestation (Nicolaides et al. 1986). By comparison, some small-for-gestational age fetuses are hypoxemic, hypercapnic, and acidotic (Nicolaides et al. 1989). Recent measurements on young (approx. 80-day old) chronically catheterized fetal sheep (term approximately 145 days), suggests that a similar gradual decline in fetal PaO_2 also occurs in this species during pregnancy (J. E. Patrick, personal communication). Superimposed on the changes in basal PaO_2 during pregnancy are decreases of 3–5 mmHg associated with episodes of uterine contractility described in the sheep as contractures (Nathanielsz et al. 1977). We showed that it was possible to mimic experimentally the contracture type of uterine activity pattern by administering oxytocin to the mothers (Lye et al. 1984). There was a direct relationship between the change in mean intrauterine pressure in response to exogenous oxytocin and the maximum decrease in PaO_2. Further, these changes were associated with a rapid increase in the concentration of adrenocorticotrophic hormone (ACTH) in the fetal circulation (Lye et al. 1985). It was speculated that increases in uterine activity and/or transient decreases in fetal PaO_2 might contribute to short-term fluctuations in plasma ACTH in fetal sheep. In turn, pulsatile changes in fetal pituitary and adrenal hormone output may contribute to normal maturation of this axis and parturition, or precocious activation may lead to preterm labor.

Acute or Short-Term Alterations in Fetal PaO_2

Short-term hypoxemia in the sheep fetus is known to effect a variety of endocrine responses, such as increases in: ACTH (Boddy et al. 1974), arginine vasopressin (AVP) (Rurak 1978; Stark et al. 1982), and cortisol (Boddy et al. 1974). In many of these studies, however, the fetal hypoxemia is associated with acidemia, and the fetal endocrine responses to different levels of hypoxemia, in the absence of acidemia, are not well established. To examine this issue, Akagi and Challis (1990) produced two levels of hypoxemia (PaO_2 change approximately 4–5 mmHg and approx. 8 mmHg), in the absence of acidemia, by altering the percentage of oxygen in the maternal inspired air. Experiments were performed on sheep at approximately 110–115 days, 125–129 days, and at more than 134 days, within 7 days of spontaneous labor. At each age, a threshold for endocrine responsiveness, measured as changes in plasma ACTH, AVP, and cortisol, was established between 4–5 and 8 mmHg. The ACTH response was significant at 110–115 days, but less than that at 125 days, although similar to that at term. In contrast, the AVP response was similar at all three gestational ages. Exogenous cortisol infusion blocked the response in ACTH, but not in AVP to moderate hypoxemia in fetuses at days 123–128, suggesting differential control mechanisms, or different sites of production for these two peptides.

It was suggested that a threshold of PaO_2 change between 4.5 and 8 mmHg exists in fetal sheep. The similar ACTH response at days 106–117, and at term, is likely due to relative immaturity of the hypothalamic-pituitary-adrenal (HPA) axis in the younger fetuses; and to glucocorticoid-mediated negative feedback in the older fetuses, where basal and stimulated plasma cortisol concentrations are higher (Akagi and Challis 1990; Challis and Brooks 1989). Interestingly there was a similar threshold (between 4.5 and 8 mmHg PaO_2 drop) for the stimulus, and a decreased incidence of rapid eye movements and breathing movements in fetuses at 125–129 days; but in term fetuses only fetal breathing, and not the incidence of eye movements, decreased with the induced change in PaO_2.

In related studies we sought changes in fetal plasma prostaglandin E_2 (PGE_2) concentrations during induced hypoxemia, since elevations in fetal plasma PGE_2 by exogenous PGE_2 infusion are known to stimulate fetal ACTH and cortisol release. Moreover, the prepartum rise in fetal plasma PaO_2 concentration is related temporally to that of cortisol (Thorburn et al. 1988; Challis and Brooks 1989). However, we were unable to demonstrate significant changes in fetal plasma PGE_2 concentrations in response to fetal hypoxemia maintained for 2 h as a result of maternal hypoxemia. This is in contrast to the stimulation of fetal PGE_2 during hypoglycemia (Thorburn et al. 1988), or during hypoxemia associated with reduced uteroplacental blood flow (see below).

Fetal β-endorphin concentrations are also elevated in hypoxemia (Stark et al. 1982), and we examined the possibility that alterations in endogenous opiates might affect fetal HPA activation. Administration of a synthetic enkephalin analogue stimulated ACTH release in fetal sheep at more than 125 days gestation, but not earlier, and the effect was blocked by concurrently infused naloxone. The opioid effect is likely at the hypothalamic rather than pituitary level, since [^{3}H]naloxone binding of high affinity was detected during the last 30 days of pregnancy in hypothalamic, but not in fetal pituitary membrane preparations (Yang and Challis 1991). In addition, there was a significant decrease in the concentration of ACTH in the fetal circulation during exogenous naloxone infusion to fetuses within 4 days of spontaneous labor (Brooks and Challis 1991). This finding implies a role for endogenous opioids in the tonic stimulation of basal ACTH release. Thus, hypoxemia may precociously accelerate normal activational processes through neuroendocrine mechanisms involving corticotropin-releasing hormone (CRH) and/or endogenous opioids, and through these pathways contribute to the stimulus to term and preterm labor.

Chronic Hypoxemia

In acute hypoxemia, the ACTH and cortisol response of the fetus is transient, and generally persists only for the duration of the hypoxemic insult. For acute hypoxemia to precipitate increased PG production and birth would require either repeated hypoxic insults or, possibly, imposition of an acute hypoxemic stimulus onto an endocrine baseline associated with a state of chronic hypoxemia. In the latter situation there may already be some increased gene expression for steroidogenic enzymes in the fetal adrenal gland, and "activation" of fetal adrenal function. We have examined the fetal endocrine responses to longer-term hypoxemia, caused by a reduction in uteroplacental blood blow. This experimental model allows separation of fetal responses from any effects modulated through the mother, and permits a study of sustained hypoxemia over 24–48 h.

Uterine blood flow was reduced by means of an adjustable teflon clamp placed around the maternal common iliac artery, and fetal oxygen saturation (SaO_2) was reduced by about 50 % from control. There was a transient acidemia at 1–2 h, which had corrected by 8 h. Initially we showed that reductions of uterine blood flow resulted in a profound rise in fetal plasma ACTH concentrations, by 50-fold, to peak values at 2 h (Challis et al. 1989). Fetal ACTH concentrations then decreased, and by 16 h were not statistically different from values in control animals, despite the sustained hypoxemia. The immunoreactive ACTH was identified by using reverse phase high-pressure liquid chromatogrpahy (HPLC) as comprising predominantly $ACTH_{1-39}$. We obtained no evidence for the secretion of large molecular weight post-translational products of pro-opiomelanocortin (POMC), although this possibility could not be excluded entirely, since we may have failed to detect such peptides with the specific antibody used in our radioimmunoassay system. More recently we found that the elevation in ACTH is preceded by a sharp increase in the concentration of CRH in the fetal circulation within 1 h of reducing uterine blood flow (Sue-Tang et al. 1990). It is possible that this CRH is of placental, rather than fetal hypothalamic origin (see below). Interestingly, fetal CRH is unchanged in fetal hypoxemia caused by maternal hypoxemia, implying that either the reduction of uterine blood flow, or the transient acidemia associated with this manipulation, precipitates the response (Sue-Tang et al. 1990). Similarly, fetal plasma PGE_2 concentrations are elevated by 4 h, and remain so for at least 24 h after reducing uterine blood flow (G. D. Thorburn and S. Hooper unpublished observations; Sue-Tang et al. 1990). No change in fetal plasma PGE_2 was seen after maternal hypoxemia, although the sampling duration of those experiments may have been insufficient to detect increased PG output.

During sustained (48 h) hypoxemia there is activation of fetal adrenal function (Challis et al. 1989). Fetal plasma cortisol concentrations showed an initial elevation, then decreased slightly, but rose progressively after 24 h, even without further change in fetal plasma ACTH. This response may mirror the activation of fetal adrenal function found after exogenous ACTH administration to the fetus (Challis and Brooks 1989), and is associated with enhanced fetal adrenal ACTH-receptor coupling and steroidogenic activity. Fetal adrenal blood flow is also elevated during either short-term hypoxemia resulting from maternal hypoxemia (Challis et al. 1986), or in sustained fetal hypoxemia with reduced uteroplacental perfusion (Challis et al. 1989). The initial rise in adrenal blood flow is associated temporally with the elevation in fetal plasma ACTH. However, the factors responsible for the sustained blood flow are unclear. It is possible that local (intra-adrenal) mechanisms are activated in response to the initial elevation in ACTH. Such mechanisms could include activation of fetal adrenal eicosanoid synthesis. Recently we found that the endothelial cells of many of the subcapsular blood vessels, that may regulate blood flow to the adrenal cortex and medulla, stained positive for PG synthase (PGHS, cyclo-oxygenase) using immunocytochemistry. Alterations in PG production, or in the synthesis of other vasoactive agents, may explain the enhanced blood flow response. In addition, such agents may contribute to increased steroidogenesis because PGs can stimulate fetal adrenal cortisol output directly (see Thorburn et al. 1988; Challis and Brooks 1989).

Because the initial endocrine responses to sustained hypoxemia occurred during an associated acidemia (Challis et al. 1989), we examined the effects on plasma hormones of infusing lactate directly into the fetus, in amounts that reduced fetal pH to less than 7.2. There was no significant change in either fetal ACTH or cortisol during lactate infusion. Similarly, Jackson et al. (1989) found no effect of lactate on fetal plasma catecholamines. In related experiments in which fetuses were made progressively hypoxemic on successive days, we found no significant change in fetal plasma ACTH until after fetal acidemia had developed (B. S. Richardson, J. Homan, J. R. G. Challis, unpublished observations). Thus, the relative importance of hypoxemia and acidemia remains to be clarified.

The importance of Placental CRH and ACTH

Maternal plasma CRH concentrations are elevated precociously in human pregnancy which is complicated by pregnancy-induced hypertension or preterm labor (see Challis and Brooks 1989). It is now clear that the source of this CRH is the placental trophoblasts, and that placental CRH mRNA, and peptide content, rise progressively throughout gestation (Frim et al. 1988). The human fetal membranes also produce CRH. Glucocorticoids stimulate CRH mRNA levels and CRH peptide output by cells from the placenta, or by membranes maintained in short-term tissue culture in a paradoxical fashion (Robinson et al. 1988; S. A. Jones et al. 1989); but there is little information on the regulation of CRH output from the placenta in vivo. C. T. Jones et al. (1988) reported secretion of immunoreactive CRH into the maternal and fetal circulation during maternal catecholamine infusion or manual occlusion of uterine blood flow. These observations support the suggestion of Sue-Tang et al. (1990), that the substantial rise in fetal plasma CRH during hypoxemia – associated with restriction of uteroplacental blood flow – likely reflects placental output. We have suggested elsewhere that the production of CRH and PGE_2 by the placenta might represent an initial compensatory response to reduced perfusion, since CRH is a vasodilator in certain vascular beds (Challis and Hooper 1989). If blood flow and fetal oxygenation are not restored, then continued secretion of CRH and PGE_2 into the fetal compartment activates the HPA axis, leading to preterm labor as an escape mechanism. Since, at least in the human, elevated fetal glucocorticoids would stimulate rather than inhibit placental CRH output, a positive feedforward cascade loop is established. The loop is broken at the time of delivery. Since

CRH may also stimulate PG production in the placenta and fetal membranes (Jones and Challis 1989), the possibility exists to accelerate further this sequence. Recent work has suggested that part of the action of CRH on placental PG output may be mediated by ACTH. The POMC gene is expressed in human placenta and membranes, and the content of ACTH in sheep chorion increases substantially at term pregnancy (R. A. Jacobs and J. R. G. Challis unpublished observations). Studies are required to determine whether acute and chronic hypoxemia in the presence or absence of altered uteroplacental blood flow alters gene expression for POMC and pre-pro CRH, or for enzymes on the PG biosynthetic pathway. Such measurements will indicate whether the effects of hypoxemia in provoking preterm birth or contributing to term labor are mediated through intrauterine PG synthesis directly, or whether they involve enhanced PG output resulting from HPAP activation.

Acknowledgements. Work in the authors' laboratory was supported by the Canadian Medical Research Council (Group Grant in Fetal and Neonatal Health and Development), and by the St. Joseph's Health Centre Foundation, and the Variety Club of Ontario.

References

Akagi K, Challis JRG (1990) Threshold of hormonal and biophysical responses to acute hypoxemia in fetal sheep at different gestational ages. Can J Physiol Pharmacol 68:540–555

Boddy K, Jones CT, Mantell C, Radcliffe JG, Robinson JS (1974) Changes in plasma ACTH and corticosteroid of the maternal and fetal sheep during hypoxia. Endocrinology 94:588–591

Brooks AN, Challis JRG (1991) Effects of naloxone on the preparturient increase in ACTH and cortisol in fetal sheep. J Neuroendocrinol 3: 419–424

Challis JRG, Brooks AN (1989) Maturation and activation of hypothalamic-pituitary-adrenal function in fetal sheep. Endocr Rev 10:182–204

Challis JRG, Hooper S (1989) Birth: outcome of a positive cascade. In: Bailliere's clinical endocrinology and metabolism 3. Saunderns, New York, pp 781–793

Challis JRG, Richardson BS, Rurak D, Wlodek ME, Patrick JE (1986) Plasma adrenocorticotropic hormone and cortisol and adrenal blood flow during sustained hypoxemia in fetal sheep. Am J Obstet Gynecol 155:1332–1336

Challis JRG, Fraher L, Oosterhuis J, White SE, Bocking AD (1989) Fetal and maternal endocrine responses to prolonged reductions in uterine blood flow in pregnant sheep. Am J Obstet Gynecol 160:926–932

Frim DM, Emanuel RL, Robinson BE, Smas CM, Adler GK, Majzoub JA (1988) Characterization and gestational regulation of corticotropin-releasing hormone messenger RNA in human placenta. J Clin Invest 82:287–292

Jackson BT, Cohn HE, Morrison SH, Baker RM, Piasecki GJ (1989) Effect of lactic acidemia on fetal adrenal secretion of norepinephrine (NE) and epinephrine (E). Proceedings of the International Union of Physiological Sciences Satellite, Fetal and Neonatal Physiology, Stockholm, July 1989, p B2

Jones CT, Gu W, Parer JT (1988) Production of CRH-like immunoreactivity by the placenta in response to alteration in uterine blood flow in pregnant sheep. In: Jones CT (ed) Fetal and neonatal development. Perinatology Press, Ithaca, pp 107–110

Jones SA, Challis JRG (1989) Local stimulation of prostaglandin production by corticotropin-releasing hormone in human fetal membranes and placenta. Biochem Biophys Res Commun 159:192–199

Jones SA, Brooks AN, Challis JRG (1989) Steroids modulate corticotropin-releasing hormone production in human fetal membranes and placenta. J Clin Endocrinol Metab 68:825–830

Lye SJ, Wlodek ME, Challis JRG (1984) Relation between fetal arterial PO_2 and oxytocin-induced uterine contractions in pregnant sheep. Can J Physiol Pharmacol 62:1337–1340

Lye SJ, Wlodek ME, Challis JRG (1985) Possible role of uterine contractions in the short-term fluctuations of plasma ACTH concentration in fetal sheep. J Endocrinol 106:R9–R11

Nathanielsz PW, Jack PMB, Krane EJ, Thomas AL, Satter S, Rees LJ (1977) The role and regulation of corticotropin in the fetal sheep. In: O'Connor M (ed) The fetus and birth. Elsevier, Amsterdam, pp 73–89 (Ciba Found Symp 47)

Nicolaides KH, Soothill PW, Rodeck CH, Campbell S (1986) Ultrasound guided sampling of umbilical cord and placental blood to assess fetal well being. Lancet i:1065–1067

Nicolaides KH, Economides DL, Soothill PW (1989) Blood gases, pH and lactate in appropriate and small-for-gestational-age fetuses. Am J Obstet Gynecol 161: 996–1001

Robinson BG, Emanuel RL, Frim DM, Majzoub JA (1988) Glucocorticoid stimulates expression of corticotropin-releasing hormone gene in human placenta. Proc Natl Acad Sci USA 85:5244–5248

Rurak DW (1978) Plasma vasopressin levels during hypoxemia and the cardiovascular effects of exogenous vasopressin in foetal and adult sheep. J Physiol (Lond) 277:341–357

Stark RI, Wardlaw SL, Daniel SS, Husain MK, Sanocka VM, James SL, Van de Viele RL (1982) Vasopressin secretion induced by hypoxia in sheep: developmental changes and relationship to β-endorphin release. Am J Obstet Gynecol 143:204–215

Sue-Tang A, Brooks AN, Hooper S, White S, Jacobs R, Bocking AD, Challis JRG (1990) Increased circulating corticotropin-releasing hormone (CRH) occurs during fetal hypoxemia with reduced utero-placental blood flow (UBF), but not as a result of maternal hypoxemia. Soc Gynecol Invest, St. Louis MO., abstr. 3/9

Thorburn GD, Hooper SB, Rice GE, Fowden AL (1988) Luteal regression and parturition: a comparison in the onset of labor: cellular and integrative mechanisms. In: McNellis D, Challis J, MacDonald P, Nathanielsz P, Roberts J (eds) The onset of Labor: cellular and integrative mechanisms. Perinatology Press, Ithaca, pp 185–206

Yang K, Challis JRG (1991) Opioid receptors are present in the hypothalamus but not detectable in the anterior pituitary of the developing ovine fetus. Neuroendocrinology 53: 623–627

Regional Abnormalities of Cerebral Energy Metabolism Following Birth Asphyxia

N. M. Bolas, B. Rajagopalan, P. Hope, J. Moorcraft,
N. K. Ives, and G. K. Radda

Introduction

The technique of ^{31}P nuclear magnetic resonance spectroscopy (^{31}P MRS) has a number of unique features that make it particularly suitable for the study of abnormalities of cellular metabolism in vivo. The information derived from a ^{31}P MRS investigation relates directly to the intracellular high energy phosphorus metabolites adenosine triphosphate (ATP) and phosphocreatine (PCr), to the inorganic phosphate (P_i) produced by breakdown of the high-energy compounds, and to the intracellular pH (pHi), which changes as a result of anaerobic glycolysis leading to lactate production. The technique is non-invasive, uses no ionising radiation, and is silent, making it particularly suitable for investigations of vulnerable patients such as the newborn for whom invasive procedures would be unduly traumatic, and for the study of inaccessible organs such as the brain.

Despite advances in patient monitoring, brain damage due to birth asphyxia remains a major cause of mortality and disability. About 0.8 % of all babies are asphyxiated to some extent at birth and require special treatment, and about one in ten of these will go on to develop a severe handicap or to die. There is currently no reliable method for assessing prognosis within the first few days of life, yet if modern intensive therapy is not to lead simply to an increase in the survival rate of severely disabled infants such an assessment is of vital importance.

Previous studies using ^{31}P MRS to investigate cerebral energy metabolism following birth asphyxia (Hope et al. 1984; Younkin et al. 1984; Laptook et al. 1980) have used spectra acquired from a large and poorly defined volume of brain tissue. Though differences in energy metabolites have been clearly seen, the specificity and sensitivity for subsequent development of neurological deficits has been disappointingly low. However we know, from post-mortem examination, that the brain damage is regional. We have therefore used a technique for obtaining spatially localised ^{31}P magnetic resonsnce spectra to investigate regional abnormalities of energy metabolism in asphyxiated infants.

Method

Principles of ^{31}P MRS

Certain atomic nuclei, including those of the biologically important and naturally abundant isotopes ^{1}H and ^{31}P, have a magnetic moment and the quantised property of spin. When placed in a magnetic field the magnetic moment of such nuclei will precess, like a gyroscope precesses when under the influence of a gravitational field. The frequency of precession (the Larmor frequency ω) is determined by the gyromagnetic ratio (γ), which is a property of the nucleus, and the precise magnetic field strength (B) :$\omega = \gamma B$. The precise magnetic field experienced by the nucleus depends on both the external applied field and the chemical environment. Thus nuclei from different chemical species will precess at slightly different frequencies, and can be distinguished. This very small frequency offset is called chemical shift, and is quantified in the relative units of parts per million (ppm):

$$((\omega - \omega_0) \times 10^6)/\omega_0$$

where ω_0 is the frequency of the signal from a reference compound, usually PCr in in vivo biological spectra.

At equilibrium, the net magnetisation from all the nuclei in a sample lies along the direction of the external applied magnetic field, and the nuclear precession cannot be detected. However if a second radiofrequency field (B_1), oscillating at the Larmor frequency, is applied to the sample for a short period, the nuclei precess in a complex manner about both the static magnetic field (B_0) and the effectively rotating field (B_1). After an appropriate time, the net magnetisation will have turned through an angle of 90° to lie perpendicular to the B_0 field. In this condition the rotating magnetisation can be detected as it will induce a voltage in a coil surrounding the sample. The oscillations continue only for a short time, until the nuclei relax. The voltage is amplified and digitised by the spectrometer, and the signal converted from ascillations in time to a frequency spectrum by a Fourier transform. Since the signal obtained is very weak, the experiment is repeated many times and the data added together.

In the simplest form of the magnetic resonance experiment, signal comes from all nuclei close to the coil (or coils). By using a small coil placed over the surface of the tissue of interest, signal can be collected largely from that tissue. While this is adequate localisation for tissues such as muscle which lie close to the surface, it is impossible to separate signals at different depths beneath the coil with such a simple system. A number of different methods have been developed to obtain better spatial localisation. These depend on one of two principles. For one set of techniques the external magnetic field (B_0) is altered so that nuclei in different places resonate at different frequencies. For the other, use is made of the fact that the applied oscillating field (B_1) gets weaker with increasing distance from the transmitter coil. The method we have applied in this study is that of phase-modulated rotating frame imaging, which belongs to the latter class. It has the following features which make it particularly suitable for the study of spatial heterogeneity in neonates:

1. It requires minimal hardware, as no special field gradients are used.
2. It is silent.
3. Signal is obtained from discs of tissue parallel to the surface, with a resolution of about 1 cm in thickness. This fits the biological structure of the brain better than other shapes, such as cubes or rectangular blocks, which are obtained using gradient techniques.
4. The result is a two-dimensional "image" with spectra at different depths resolved in one direction, and chemical shift in the other direction. This makes it easy to compare different regions of the same tissue, in contrast to most of the gradient-based techniques in which signal is obtained from only one region at a time.

The method of phase-modulated rotating frame imaging is based on the following principles. As previously mentioned, application of a radiofrequency field causes the nuclei to precess not only about the static B_0 field, but also about the rotating B_1 field. For ease of visualisation one can effectively eliminate the B_0 field by considering what happens from a viewpoint which is also rotating at the Larmor frequency. In this frame of reference, the rotating frame, only the applied B_1 field need be considered, and this is apparently static. When the B_1 field is applied, nuclei will start to precess about it, at a rate that depends on the field strength. Thus close to the coil the precession will be faster, and further away the precession will be slower. If the pulse is applied for a certain length of time then nuclei close to the coil will have turned through a larger angle than those further away. The amplitude of the signal induced in the receiver coil depends on the sine of the angle through which the nuclei turn. So by applying a series of pulses of incrementing duration, the signal originating from a part of the sample close to the coil will be modulated rapidly, while that from material further away will be modulated more slowly. A second Fourier transform, this time using pulse length as the time dimension, will result in signal close to the coil appearing at the high frequency end, and signal from further away appearing at the low frequency end of the spatial spectrum.

The experiment as so far described is inherently inefficient, as no signal is collected from nuclei that happen to have turned through multiples of 180°. By applying a second orthogonal 90° pulse after the incremental pulse all the signal can be detected, and the incremental pulses will result in a phase modulation rather than an amplitude modulation of the received signal. For further description of the precise implementation, see Blackledge et al. (1987).

Application of ^{31}P MRS

The infant was positioned with the right temporoparietal region of the brain over a double surface coil consisting of a 6.5-cm diameter receiver coil and an offset 1.5-cm diameter transmitter coil (Styles 1988). The coil was tuned, and located at the centre of a 1.9 T 60-cm bore magnet. The magnetic field homogeneity was optimised using the ^{1}H signal from water, and a ^{31}P spectrum was collected from the whole region under the coil (pulse length 275 μs; interpulse delay 3.012 s;128 transients). The rotating frame image data were then collected (incremental and phase encoding pulse lengths 280–300 μs; interpulse delay 3.012 s; 8–12 increments; 16–32 scans per increment). Data were collected and processed using a Bruker Biospec spectrometer. Free induction decays were filtered using an exponential line broadening of 15 Hz, and the broad component in the spectrum due to bone and immobile phospholipids was removed by "profile correction" (Gordon et al. 1982) with a linewidth factor of 400 and an amplitude factor of 4, before Fourier transformation. The whole investigation took about 45 min.

Asphyxiated and Control Infants

Asphyxiated infants were among those admitted to the Special Care Baby Unit between January 1986 and May 1989. The severity of asphyxia was assessed using the Sarnat encephalopathy score, except that grade 2 was subdivided into grade 2a (clinical improvement before 1 week of age) and 2b (no improvement before 1 week). Such encephalopathy grades were generally assigned after the magnetic resonance investigation. Results were then considered in three classes; normal (no asphyxia), mild (encephalopathy score 0, 1, 2a) and severe (encephalopathy score 2b, 3). All babies were studied with the informed consent of the parents, and the studies were approved by the local ethics committee. Table 1 summarises the clinical information. During the investigation breathing rate and electrocardiograp-

Table 1. Clinical information obtained from studies in asphyxiated and control infants

	Degree of asphyxia at birth		
	Normal controls	Mild	Severe
n	4	16	10
Gestation (weeks)	39 ± 2	39 ± 2	40 ± 2
Birth weight (g)	3247 ± 653	2885 ± 625	3380 ± 634
Age at MRS (days)	6.5 ± 5.8	3.3 ± 1.6	6.5 ± 4.7
Apgar 1 min	9 ± 0	3 ± 1.9	1.9 ± 2
Apgar 5 min	10 ± 0	6.8 ± 1.8	4.3 ± 3.2

All values are mean ± s.d.

hic (ECG) data were monitored, and a paediatrician accompanied the baby at all times. Babies were not sedated, but were studied during natural sleep. Due to the sensitivity of the localisation method to movement, only data from babies who remained still throughout the study have been included in the results.

Results and Discussion

A typical non-localised spectrum from a mildly asphyxiated infant is shown in Fig. 1. A dramatic rise in the peak due to P_i in comparison to the other peaks in the spectrum was seen in babies following severe asphyxia. We have not attempted to determine absolute concentrations of metabolites in these experiments, but the ratios of peak areas, and pHi, from non-localised spectra are presented in Table 2.

The energy metabolism in the brain can be summarised by the following diagram:

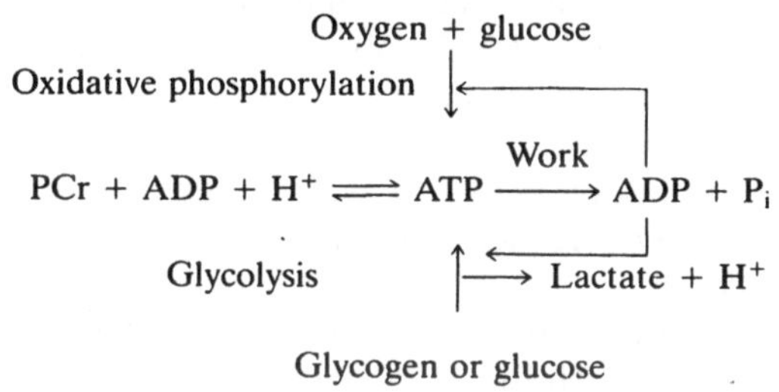

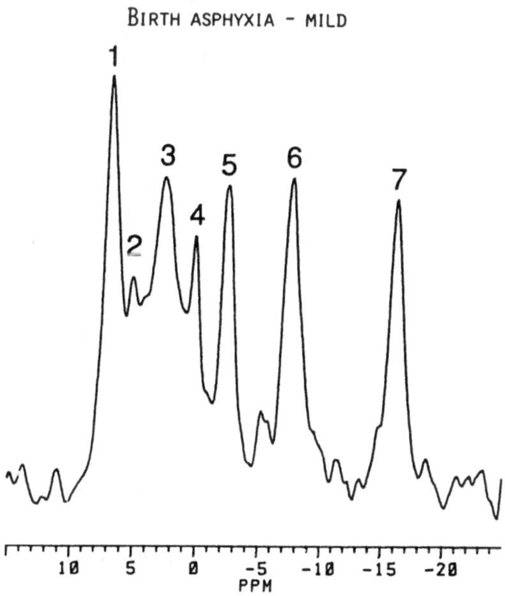

Fig. 1. Non-localised ^{31}P MRS spectrum from a mildly asphyxiated infant. Peak assignments are *1* phosphomonoesters (largely phosphoethanolamine), *2* P_1, *3* phosphodiesters (immobile components of membranes), *4* PCr, *5* γ phosphate of ATP, *6* α phosphate of ATP, and *7* β phosphate of ATP

Table 2. Ratios of peak areas and pHi from non-localised spectra

	Degree of asphyxia at birth		
	Normal controls	Mild	Severe
P_i/PCr	0.54 ± 0.35	0.72 ± 0.25	1.99 ± 1.46*
P_i/ATP	0.28 ± 0.21	0.34 ± 0.10	0.63 ± 0.34

* $p < 0.01$ mild vs severe (Mann-Whitney)

ATP is hydrolysed to adenosine diphosphate (ADP) and P_i with the release of energy for maintenance of ion gradients, etc. Normally ATP is resynthesised from ADP by oxidative phosphorylation, in a very tightly regulated manner so that the concentration of ATP remains constant. In the absence of oxygen, ATP is maintained either by glycolysis, or by the breakdown of PCr. Both of these pathways are amenable to investigation by ^{31}P MRS; PCr breakdown and the consequent increase in P_i are detected by the relative changes in the corresponding peak areas, and glycolysis results in the production of lactate and a decrease in pHi, which is detected as a shift in the position of the P_i peak.

From this outline, we would expect the PCr/P_i ratio to be the most sensitive indicator of energetic stress, with PCr/ATP, P_i/ATP, and pHi being alternative indicators. The results in Table 2 confirm that there are significant changes in these parameters correlating with encephalopathy. However there is a large variance, particularly in the results from normal and mildly asphyxiated infants.

Figure 2 illustrates the phase-modulated rotating frame imaging results obtained from a mildly asphyxiated infant. Comparable results from a severely asphyxiated infant are shown in Fig. 3. The global patterns of increased P_i and decreased PCr are reflected in these images. The P_i peak seen in the severely asphyxiated infant spectra is very much larger than that in the mild example, and progresses much more deeply into the brain tissue. Conversely PCr is very much reduced, and is localised only to the superficial tissue, in the severely asphyxiated infant. The progressive increase in P_i with increasing depth is characteristic of the more severe grades of asphyxia (Table 3).

We have seen (results not shown) that the changes with asphyxia in grade 2a are confined to the deepest tissue layers, while in the severe grades 2b and 3, tissue damage is apparent also in the middle layer, but not at a superficial level. The clinically most important distinction is between grade 2a and 2b, since this subdivision can only be assigned retrospectively and relates closely to the future progression of the infant.

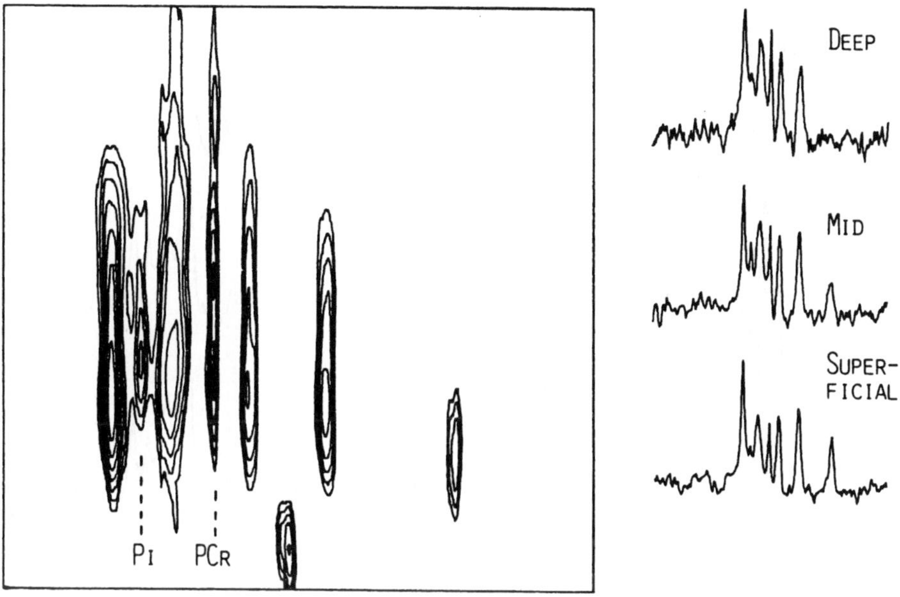

Fig. 2. Phase-modulated rotating frame image from a mildly asphyxiated infant. The contour plot (*left*) shows chemical shift (*horizontal axis*) against depth (*vertical axis*). The contour shapes progress from close to the coil (*bottom*) to a depth of about 4 cm (*top*). The contour lines show changes in peak height. Careful inspection of the PCr peak in this figure shows a small dip in peak height just below the centre of the image, due to local heterogeneity. The small peak at the bottom of the image is from a phantom placed within the receiver coil. Individual spectra from three depths corresponding to horizontal lines drawn through the contour plot, are shown on the right. Here the local heterogeneity can again be seen by comparing the heights of the PCr and P_i peaks at the three depths

Table 3. Relationship between P_i/ATP ratio and degree of asphyxia at various depths

	Degree of asphyxia	
p_i/ATP ratio	Mild	Severe
Superficial	0.30 ± 0.20	0.49 ± 0.17
2 cm depth	0.39 ± 0.19	0.66 ± 0.31
3 cm depth	0.36 ± 0.22	0.95 ± 0.39

The image from the mildly asphyxiated infant in Fig. 2 also clearly shows heterogeneity within the brain. In this infant PCr is seen to decrease in the mid region, at approximately 2 cm depth, and to increase again in deeper tissue. Such patterns of heterogeneity have been seen in many of the normal

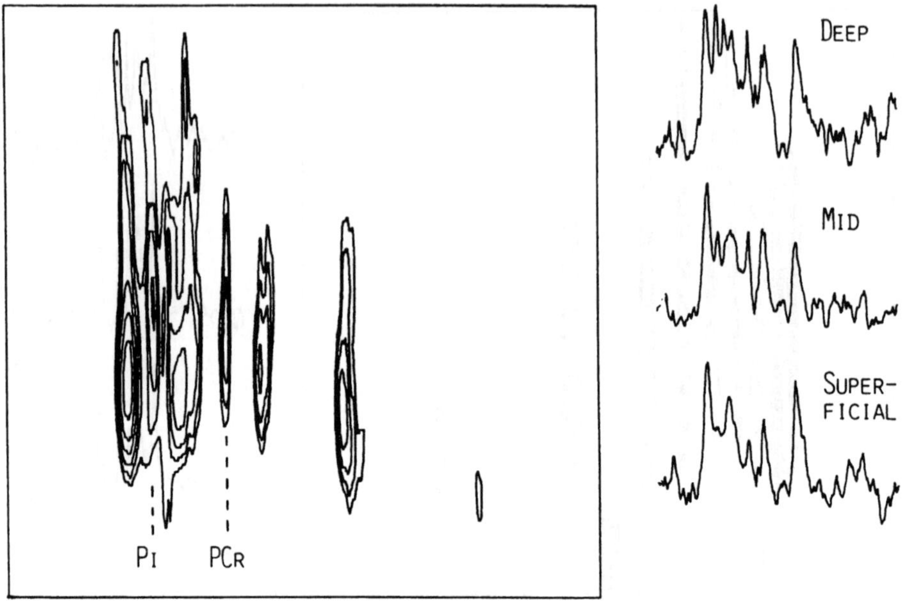

Fig. 3. Phase-modulated rotating frame image from a severely asphyxiated infant. In comparison with Fig. 2, the increase in the P_i peak and the decrease in the PCr peak can clearly be seen. The individual rows show the progressive increase in P_i compared to the other peaks with increasing depth into the brain

and more mildly asphyxiated infant spectra, but no consistent pattern of changes is seen. The variation is much larger than can be explained simply by the amount of noise in the spectra, and must reflect true variations in the metabolite leels within the brain. The variability is greater in the superficial layers of the brain. This may explain some of the variability of the non-localised spectra, since the majority of signal in spectra obtained with a simple surface coil derives from tissue close to the surface.

Conclusions

The technique of ^{31}P MRS offers unique opportunities for the investigation of energy metabolism in vivo. Applications to the investigation of birth asphyxia show gross changes in ratios of the energy metabolites PCr, P_i and ATP, and in pHi. However spectra obtained from the whole head cannot be used to clearly predict prognosis. In an attempt to improve the sensitivity and specificity of the method, we have obtained spatially localised spectra using phase-modulated rotating frame imaging. The resulting spectra reveal that the disruption of energy metabolism following birth asphyxia is more severe in the deeper layers of brain tissue. Spectra obtained from the deeper layers offer a better chance of helping to assess prognosis. The localised spectra also show an unexpected degree of metabolic heterogeneity in normal and mildly asphyxiated infants, particularly in the superficial layers of brain tissue.

References

Blackledge MJ, Rajagopalan B, Oberhaensli RD et al. (1987) Quantitative studies of human cardiac metabolism by ^{31}P rotating frame nmr. Proc Natl Acad Sci USA 84:4283

Gordon RE, Hanley PE, Shaw D (1982) Topical magnetic resonance. Prog Nucl Magn Reson Spectrosc 15:1–47

Hope PL, Cady EB, Tofts PS et al. (1984) Cerebral energy metabolism studied with phosphorus NMR spectroscopy in normal and birth asphyxiated infants. Lancet ii:366–370

Laptook AR, Corbett RJT, Uaug R et al. (1989) Use of ^{31}P magnetic resonance spectroscopy to characterise evolving brain damage after perinatal asphyxia. Neurology 39:709–712

Styles P (1988) Passive electrical isolation of double coil probes. NMR Biomed 1:61–66

Younkin DP, Delivoria-Papadopoulos M, Leonard JC et al. (1984) Unique aspects of human newborn cerebral metabolism studied with phosphorus nuclear magnetic resonance spectroscopy. Ann Neurol 16:581–586

Fetal Heat Flux and Oxygen Tension During Asphyxia

R. Rudelstorfer

Introduction

The literal meaning of asphyxia is "not to throb" or a "stopping of the pulse". In clinical terms, asphyxia refers to the loss of consciousness as a result of too little oxygen. Today and during the intrapartum period, asphyxia is commonly referred to as intrapartum acidosis. One goal of modern obstetrics is the avoidance of such an event. Cardiotocography and intermittent scalp blood sampling are the established diagnostic procedures in this field. However, the diagnosis of intrapartum acidosis is not always reliable without too many interventions on behalf of the fetus and/or mother.

We feel that measuremtns of the fetal heat flux could improve the situation. The major reason for this is that fetal metabolic activity and heat production are more closely related tò its oxygen consumption and lactic acid production than heart rate alterations. The diagnostic principle of fetal heat flux measurements is based on the *quantitative* assessment of metabolic heat loss. Although in clinical routine heat flux measurements are performed from the presenting part of the fetus (i.e., fetal scalp), experimental studies have indicated a good correlation to total body surface heat loss (Rudelstorfer et al. 1986; Simbruner et al. 1985).

But how can intrapartum asphyxia be detected by heat flux measurements? The following short introduction to thermal homeostasis between the fetus and the mother will answer this question. During the last trimenon, the fetus has a body temperature whch is about 0.2°–0.3 °C higher than the surrounding maternal tissues (Rooth et al. 1977). Therefore, and according to the second law of thermodynamics, the fetal organism exchanges heat constantly toward the mother. During steady state conditions, there is a equilibrium between heat production and heat flux. Heat, which is generated by the metabolic activity of fetal tissues, is diffused via two sites: (a) the fetal surface by conduction and (b) the umbilical circulation and the placenta by convection. Heat convection by the umbilical circulation represents by far the major route for fetal heat loss (approximately 80 %–90 % under nonstress conditions) (Rudelstorfer et al. 1986).

Interpretation of Heat Flux Measurements

Assuming a steady heat production by the fetal metabolism and supposing that the heat removal by the umbilical circulation is reduced (e.g., cord compression), then there must be an increase of heat convection via the fetal scalp. On the other hand, assuming a steady rate of heat removal but supposing that heat production is reduced (e.g., due to fetal hypoxemia), then there must be a decrease of heat convection via the scalp. However, one should always be aware of the hypothetical nature of this concept because, in a living organism, these processes do not take place in a strictly isolated way. As a basis and for diagnostic purposes, however, it is sufficient to know that low heat flux values represent hypoxemia and high values indicate disturbances of heat diffusion (in most instances via the placenta) (Zilianti et al. 1983; Rudelstorfer et al. 1987).

Material and Methods

Selection Criteria

Out of 207 women in whom fetal heat flux had been measured from the fetal scalp during labor and delivery, a selection was made according to following criteria: single fetus, vertex presentation, maternal axillary temperature ≤37.5 °C, and postpartum determination of pH ($n = 136$) and oxygen tension (pO_2) ($n = 51$) in the umbilical artery. To classify the fetuses according to the nature of the asphyxia, four groups were formed: (a) acidotic fetuses ($n = 34$) with a pH of the umbilical artery of ≤7.19; (b) nonacidotic fetuses ($n = 102$) with a pH ≥7.20; (c) hypoxemic fetuses with a pO_2 below two standard deviations of our nonacidotic controls (i.e., $pO_2 < 10$ mmHg; $n = 7$); and (d) 44 nonhypoxemic fetuses which served as controls. Furthermore, seven fetuses were included which had died in utero earlier. They represented the extreme form of intrapartum asphyxia and served as a control value for nonmetabolic heat exchange. The study was approved by the Ethic Committee of the University of Vienna Medical School, and informed oral consent was obtained from all mothers.

Method

After the membranes had ruptured and the cervix had dilated to >3 cm, fetal scalp heat flux was mesured by means of a round heat flux transducer, 25 mm in diameter and 1 mm in thickness (model Ha 15-15-13[C], Thermonetics, San Diego, California). The transducer contains thermopiles in a thin, heat-conductive platelet and produces an electrical signal. The heat flux is expressed in units of W/m^2. The transducer was inserted either directly or using an amnioscope or a specially designed introducer and then attached to the parieto-occipital region of the fetal scalp. The signal was ampliefied and recorded on a polygraph (R 611, Beckman Instruments, Fullerton, California, or Hellige 6, Heidelberg, FRG) or on a specially adapted cardiotocograph (Hewlett-Packard, Palo Alto, California). The fetal heart rate and the uterine activity were recorded with commercially available cardiotocographs. After delivery, the umbilical cord was clamped and a segment was brought to the laboratory. Blood from the umbilical artery and vein were analyzed separately with an AVL Gascheck 845 (AVL, Graz 8010, Austria) with automatic calibration. The axillary temperature of the mother and the air temperature of the delivery room were measured with standard mercury thermometers. The heat flux recordings were analyzed for the last 60 min before delivery in 10-min intervals from artifact-free recordings. The basal heat flux values, i.e., the heat flux just before the onset of a contraction, were evaluated in this study. Mean heat flux values refer to this basal heat flux during this 60-min period. Most readings included parts of the first and second stages of delivery. Full dilatation of the cervix was usually documented directly on the chart.

Statistical Analysis

Data were reported as means ± standard deviation. For statistical evaluation, we performed an analysis of variance (ANOVA) and linear regression. Significance was set at $p < 0.05$.

Results

Heat flux values from the fetal scalp of healthy nonasphyxiated, nonacidotic fetuses were in the range of 15 ± 3.5 W/m^2. Significantly lower values were found in acidotic and hypoxic fetuses (Table 1). We found a significant linear correlation between the pO_2 in the umbilical artery and the mean heat flux values within the last 30 min of delivery (Fig. 1). As the time interval increased between heat flux measurements (i.e., >30 minutes before delivery) and the determination of pO_2 the relationship between these two variables diminished.

Heat flux in fetuses which had died in utero earlier showed very low values and had no uniform pattern of heat flux. There were periods of practically no heat flux whatsoever, followed by heat uptake from the mother and then reversed flow from the fetus back to the mother (Fig. 2). These changes reflected either changes in the maternal body temperature, or they were observed in association with contractions of the uterus during the second stage.

Table 1. Intrapartum asphyxia: scalp heat flux and acid base status of umbilical artery

Group	n	Heat flux (W/m^2) Mean	SD	pH (log) Mean	SD	pO_2 (mmHg) Mean	SD	pCO_2 (mmHg) Mean	SD	Base excess
Nonacidotic (pH ≥7.20)	102	15.1	4	7.27	0.04	19.4	6	46.0	12	− 6
Acidotic (pH ≤719)	34	12.0	6**	7.13	0.08*	12.9	5*	60.0	11*	−11*
Nonhypoxemic (pO2 ≥10 mmHg)	44	15.3	4	7.25	0.07	19.4	6	46.0	12	− 6
Hypoxemic (pO2 ≤9 mmHg)	7	7.7	4*	7.13	0.13*	6.9	2*	55.0	12	−12*

* $p < 0.01$
** $p < 0.04$

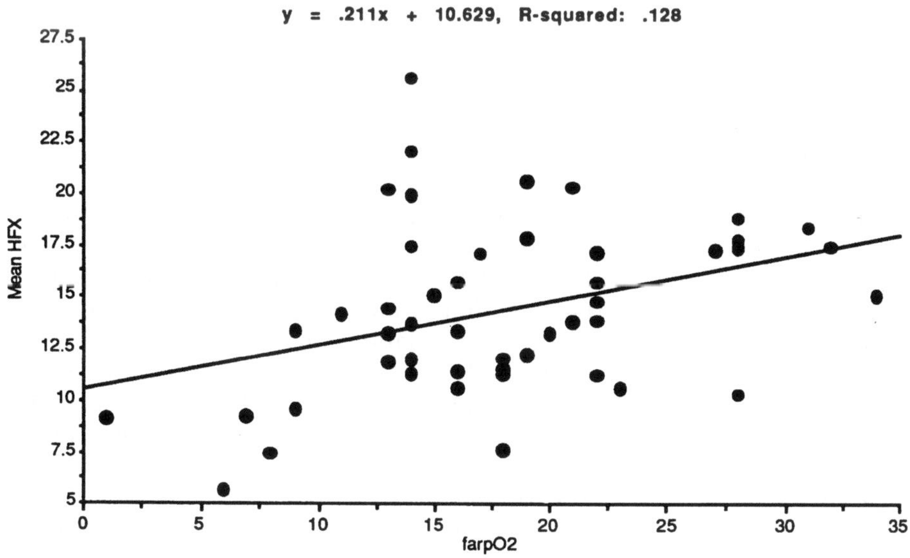

Fig. 1. Linear regression between pO_2 of the umbilical artery (*farpO2*) and heat flux values (watts/m^2) within 30 min of delivery ($r = 0.36$; $n = 51$; $p < 0.02$)

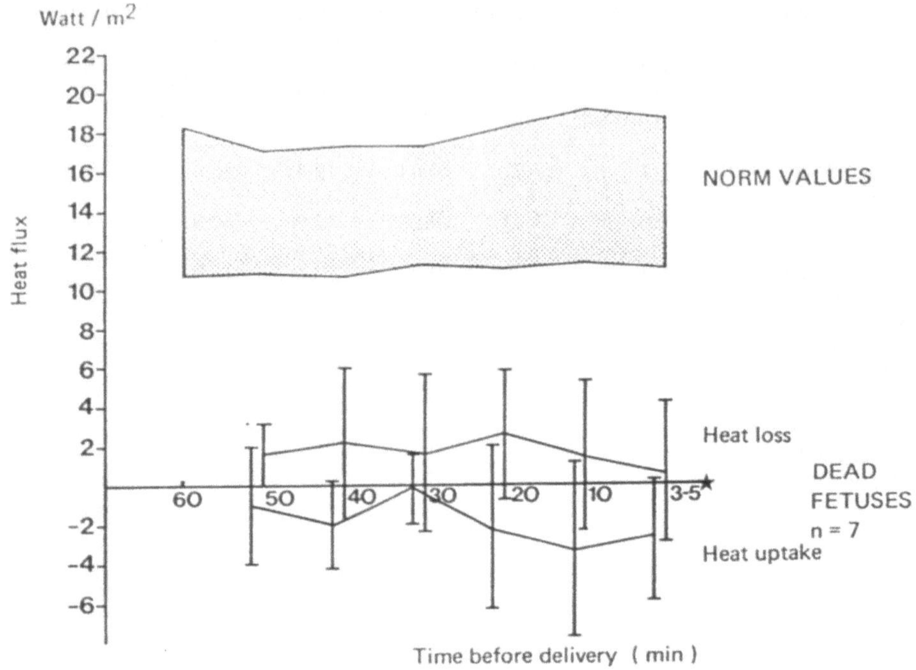

Fig. 2. "Maximum" and "minimum" heat flux in dead fetuses, which constitute a extreme form of "asphyxiated" fetuses. Note that heat uptake as well as heat loss could occur during a 10-min observation period. In summary, the heat flux approximated zero

Discussion

Heat flux measurements yielded information about the metabolic activity of the fetus which were unique and different from those obtained by cardiotocography. The amount of fetal heat loss via the scalp indicated metabolic activity as well as the current situation of placental exchange function (for heat). It seems that there is a functional relationship between pO_2 in the fetal blood and the amount of heat flux from the scalp, especially in the lower range of pO_2. Low heat flux values indicated fetuses with reduced metabolic activity, and these fetuses could also be identified as asphyxiated according to intrapartum acidosis and/or hypoxemia. From a practical viewpoint, heat flux measurements are less invasive than scalp blood sampling procedures (Rudelstorfer et al. 1987). The heat flux method does not require ruptured membranes and maintenance of a sophisticated equipment. We are convinced that at some time in the future this method could render a desirable service for detecting intrapartum asphyxia.

References

Rooth G, Huch A, Huch R, Peltonen R (1977) Fetal-maternal differences during labor. Contrib Gynecol Obstet 3:54–62

Rudelstorfer R, Tabsh K, Khoury A et al. (1986) Heat flux and oxygen consumption of the pregnant uterus. Am J Obstet Gynecol 154:462–470

Rudelstorfer R, Simbruner G, Sharma V, Janisch H (1987) Scalp heat flux and its relationship to scalp blood pH of the fetus. Am J Obstet Gynecol 157:372

Simbruner G, Weninger M, Popow C, Hergoldt WJ (1985) Regional heat loss in newborn infants. Part II. Heat loss in newborns with various diseases – a method of assessing local metabolism and perfusion. S Afr Med J 68:948

Zilianti M, Cabello F, Chacon NR, Rinicon CS, Calazar JR (1983) Fetal scalp temperature during labor and its relation to axid-base balance and condition of the newborn. Obstet Gynecol 61:474

Abstracts

Brain Blood Flow and the Dynamics of Circulatory Centralization During Acute Asphyxia in Intact and Chemically Sympathectomized Fetal Sheep

A. JENSEN and U. LANG

Acute fetal asphyxia caused by a reduction in uterine blood flow both increases sympathetic activity and decreases blood flow to peripheral organs, including the skin (Jensen and Berger 1991). These changes are accompanied by rapid increases in arterial blood pressure and in blood flow to central organs, i.e., brain, heart, and adrenals. The rapidity and uniformity of this peripheral vasoconstriction suggest that it may be a reflex of the sympathoneuronal system which causes these initial blood pressure and peripheral flow changes in order to maintain oxygen delivery to the brain during acute asphyxia.

To test this hypothesis and to further validate skin blood flow as an index of fetal circulatory centralization, 11 chronically prepared normoxemic unanesthetized near-term fetal sheep in utero were used (term is at 147 days). In five fetuses we studied the time course of changes in both organ blood flow (microsphere method, Jensen et al. 1987) and catecholamine concentrations using high-performance liquid chromatography with electrochemical detection (HPLC-ECD) (Jelinek and Jensen 1991) during and after arrest of uterine blood flow for 2 min, i.e., at 0, 1, 2, 3, 4, and 30 min. Secondly, to assess the influence of the sympathoneuronal system on the time course of circulatory centralization separately from the influence of the adrenal medulla thereon, we administered 6-hydroxydopamine to six other fetuses to selectively destroy their sympathetic nerve terminals, and we then examined the circulatory and hormonal responses of these sympathectomized fetuses using the same protocol. In intact fetal sheep near term, arrest of uterine blood flow for 2 min caused an increase in arterial blood pressure and a rapid circulatory centralization systemically and within the brain in favor of the heart and brain stem, respectively, and at the expense of blood flow to peripheral organs, including skin and scalp. The decreases in blood flow to the fetal skin and scalp during asphyxia and their increases after asphyxia reflected the blood flow changes in other principal peripheral organs.

In chemically sympathectomized fetuses, arterial blood pressure fell transiently at 1 min of asphyxia, and cardiac output was redistributed towards the carcass and intestinal organs at the expense of the brain stem, heart, and placenta (Figs. 1, 2).

We conclude that in fetal sheep near term, changes in blood flow to the skin and scalp during and after acute asphyxia indicate both the development of

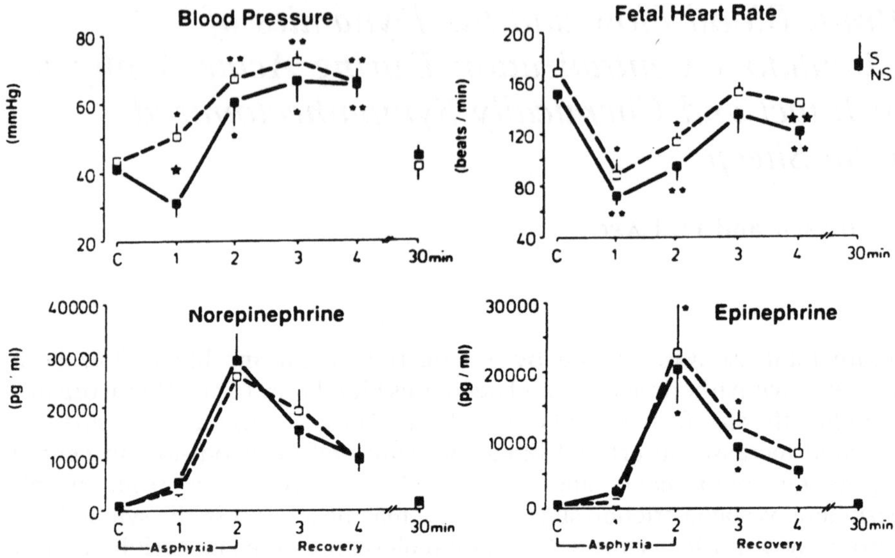

Fig. 1. Changes in intact (*NS*) and sympathectomized (*S*) fetal sheep near term before, during, and after 2 min of asphyxia

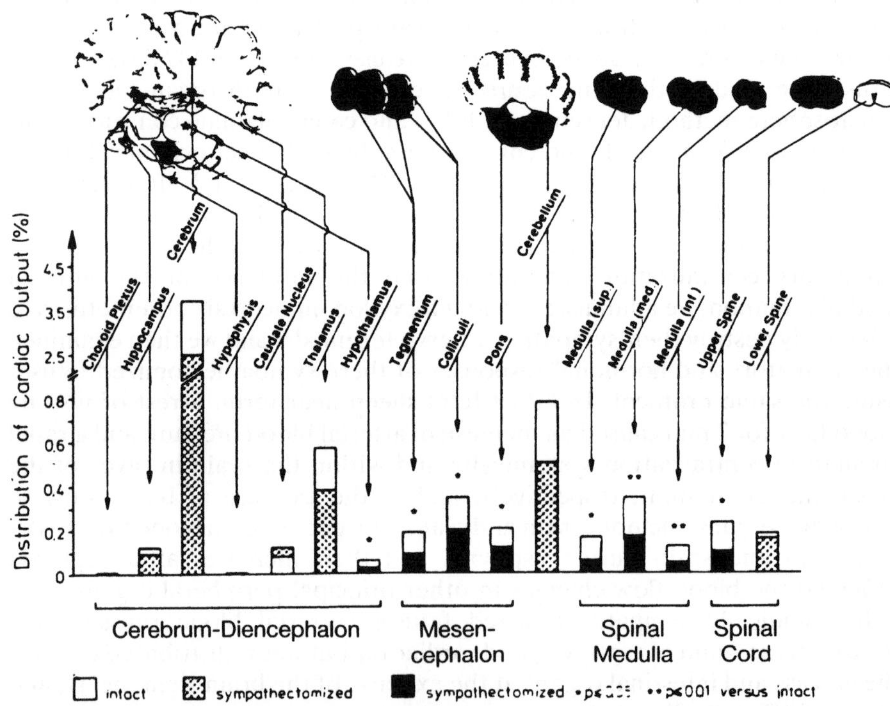

Fig. 2. Fraction of cardiac output distributed to the brain of intact and sympathectomized fetal sheep near term. *Black areas* indicate brain parts in which blood flow is significantly reduced in sympathectomized fetuses after 1 min of asphyxia

and the recovery from fetal circulatory centralization. And we further conclude from our studies on chemical sympathectomy that the initial phase of this circulatory centralization, in which blood flows to the brain stem and heart increase, is highly dependent on the sympathoneuronal system. Thus, the sympathoneuronal system is important for fetal short-term adaptation to and intact survival of asphyxia.

Acknowledgements. This work was supported by the Deutsche Forschungsgemeinschaft grant Je 108/4-1.

References

Jelinek J, Jensen A (1991) Catecholamines concentrations in plasma and organs of fetal guinea pig during normoxaemia, hypoxaemia and asphyxia. J Dev Physiol 15:145–152

Jensen A, Hohmann M, Künzel W (1987) Dynamic changes in organ blood flow and oxygen consumption during acute asphyxia in fetal sheep. J Dev Physiol 9:543–559

Jensen A, Berger R (1991) Fetal circulatory responses to oxygen lack. J Dev Physiol 16:181–207

Dynamics of Circulatory Centralization During Acute Asphyxia in Preterm and Term Fetal Sheep

A. JENSEN, U. LANG, and G. BRAEMS

In fetal sheep near term (term is at 147 days), acute asphyxia caused by reduction in uterine blood flow increases both sympathetic activity and arterial blood pressure and decreases blood flow to peripheral organs, including the skin, to maintain oxygen delivery to the brain and heart (Jensen et al. 1987, Jensen and Berger 1991). This rapid centralization depends largely on the sympathetic nervous system (Jensen and Lang 1988). Since the sympathetic nervous system matures during development, circulatory centralization may be different in younger fetuses.

To test this hypothesis and to further validate skin blood flow as an index of fetal circulatory centralization, we studied five near-term (130 ± 2 days) and seven preterm (110 ± 3 days) chronically prepared normoxemic unanesthetized fetal sheep in utero. The time course of changes in organ blood flow (microsphere method), plasma catecholamine concentrations (high-performance liquid chromatography with electrochemical detection, HPLC-ECD), blood gases, pH, arterial pressure, and heart rate were measured during and after arrest of uterine blood flow for 2 min, i.e., at 0, 1, 2, 3, 4 and 30 min. In preterm *and* near-term fetal sheep, arrest of uterine blood flow for 2 min caused a rapid circulatory centralization in favor of the brain stem and heart and at the expense of blood flow to peripheral organs, including skin and scalp. In both groups, the decreases in blood flow to the fetal skin and scalp during asphyxia and their increases after asphyxia reflected the blood flow changes in other peripheral organs. However, during asphyxia in preterm fetuses, arterial blood pressure, catecholamines, and cerebral blood flow were significantly lower than in near-term fetuses, resulting in a poorer oxygen delivery to the brain and heart in the previous group (Fig. 1). We conclude that in preterm and near-term fetal sheep, changes in blood flow to the skin and scalp during and after acute asphyxia indicate both the development of and the recovery from fetal circulatory centralization. We further conclude that in preterm fetuses this circulatory centralization is less effective. Thus, intact survival of asphyxia is less likely in preterm than in near-term fetuses.

Acknowledgements. This work was supported by DFG grant Je 108/4-2.

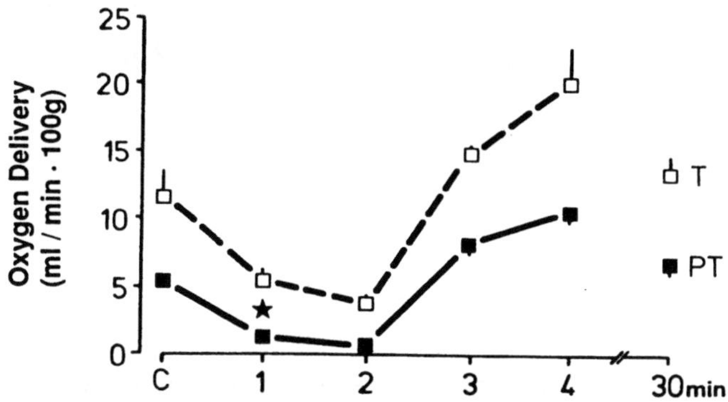

Fig. 1. Oxygen delivery to the brain of preterm (*PT*) and term (*T*) fetal sheep during and after acute asphyxia

References

Jensen A, Lang U (1988) Dynamics of circulatory centralization and release of vasoactive hormones during acute asphyxia in intact and chemically sympathectomized fetal sheep. In: Künzel W, Jensen A (eds) The endocrine control of the fetus. Physiologic and pathophysiologic aspects. Springer, Berlin, Heidelberg, New York, pp 135–149

Jensen A, Lang U, Künzel W (1987) Die Dynamik des fetalen Kreislaufschocks bei akuter Asphyxie: die Kopfhaut ist ein Schockindikator. 13th German Congress for Perinatal Medicine, Berlin, 1–4 December 1987. In: Dudenhausen JW, Saling E (eds) Perinatale Medizin, vol XII. Thieme, Stuttgart, pp 178

Jensen A, Berger R (1991) Fetal circulatory responses to oxygen lack. J Dev Physiol 16: 181–207

*Regional Brain Blood Flow and Oxygen Delivery in Guinea Pig Fetuses Near Term**

J. JELINEK, A. M. CARTER, and A. JENSEN

To examine brain blood flow and oxygen delivery to 14 brain regions in fetal guinea pigs near term, six fetuses were catheterized under general anesthesia of the dam on days 61–63 of pregnancy. In the control period, fetal arterial blood pressure was measured and a blood sample was taken from the right axillary artery to determine fetal blood gases and acid base balance. Isotope-labeled microspheres were then injected in a lateral saphenous vein, and a reference blood sample was taken from the axillary artery according to Carter and Gu (1988).

Preductal arterial oxygen saturation of hemoglobin ranged from 16%–33% (mean ± SD, 26.7 ± 6.3) and the pH from 7.08 to 7.30 (7.204 ± 0.09); oxygen content varied between 3.0 and 6.7 ml/dl (6.1 ± 1.3). Blood flows to the whole brain were 105–258 ml/min per 100 g (167 ± 58), and oxygen delivery to the brain was 6.61–10.67 ml/min per 100 g (8.00 ± 1.55). Both regional brain blood flow and oxygen delivery corresponded well to the values described for the fetal sheep (Jensen et al. 1985, 1987). They were highest in the spinal medulla and other parts of the brain stem and lowest in the cerebrum. Brain blood flow increased significantly with increasing pCO_2 and systolic blood pressure and with decreasing oxygen saturation of hemoglobin, pH, base excess, and arterial oxygen content of the blood. Regarding the slopes of the regression lines, regional brain blood flow increased most in the pons and in the spinal medulla and least in the cortex. On the other hand, except for the middle portion of the spinal medulla, there were no correlations between blood gas and acid base balance variables and the oxygen delivery to the whole brain or to various regions of the brain.

We conclude that both the amount and the distribution of brain blood flow in fetal guinea pigs near term agrees well with those of other animal species. Furthermore, there is a close relationship between brain blood flow and the fetal oxygenation, but with regional differentiation.

* This investigation was supported by the Deutsche Forschungsgemeinschaft (Je 108/4-1) and the Danish Medical Research Council

References

Carter AM, Gu W (1988) Cerebral blood flow in the fetal guinea-pig. J Dev Physiol 10:123–129

Jensen A, Künzel W, Kastendieck E (1985) Repetitive reduction of uterine blood flow and its influence on fetal transcutaneous PO_2 and cardiovascular variables. J Dev Physiol 7:75–87

Jensen A, Hohmann M, Künzel W (1987) Redistribution of fetal circulation during repeated asphyxia in sheep: effects on skin blood flow, transcutaneous PO_2, and plasma catecholamines. J Dev Physiol 9:41–55

*Plasma and Tissue Concentrations of Catecholamines During Normoxemia, Hypoxemia, and Asphyxia in Fetal Guinea Pigs Near Term**

J. Jelinek and A. Jensen

Fetal guinea pigs are frequently used for perinatal investigations, but little is known about the function of their sympathetic nervous system during hypoxemia and asphyxia. To examine the responses of the sympathetic nervous system to oxygen deficiency, we studied plasma and tissue concentration of catecholamines during normoxemia, hypoxemia, and asphyxia in 22 fetal guinea pigs near term (Jelinek and Jensen 1991). Fetal blood was obtained by cardiopuncture in utero under ketamine/xylazine anesthesia of the dam. Catecholamines were determined in plasma and tissue of 15 organs and 14 brain parts by high-performance liquid chromatography with electrochemical detection.

During normoxemia (SO_2, 54 ± 4%; pH, 7.36 ± 0.02; $n = 5$) plasma catecholamine levels were low (norepinephrine 447 ± 53; epinephrine 42 ± 12; dopamine 44 ± 6 pg/ml). During hypoxemia (SO_2, 27 ± 3%; pH, 7.32 ± 0.01; $n = 6$) ans asphyxia (SO_2, 24 ± 2%; pH, 7.23 ± 0.02; $n = 11$), tissue catecholamine concentrations changed with changing blood gases and with increasing plasma catecholamines, but in different organs in different directions. Norepinephrine concentrations increased in the skin and lung and decreased in the liver, pancreas, and scalp, and epinephrine increased in the heart, lung, liver, and scalp and decreased in the adrenal. These changes in tissue catecholamine concentrations also resulted in marked changes in the norepinephrine/epinephrine ratio in many organs in relation with changes in blood gas values and acid base balance. In the brain there were only minor changes in catecholamine concentrations but concentrations of dopamine catabolite 3,4-dihydroxyphenylacetic acid decreased in many brain parts, suggesting that the metabolism of cerebral catecholamines was affected by hypoxemia and asphyxia.

We conclude that the sympathetic nervous system of the fetal guinea pig near term is mature and that it is stimulated by reduced fetal oxygen supply. Changes in both plasma and tissue catecholamine concentrations and their interrelations may be important for vascular, metabolic, and endocrine reactions during hypoxemia and asphyxia.

---

* This investigation was supported by the Deutsche Forschungsgemeinschaft (Je 108/4-1)

Reference

Jelinek J, Jensen A (1991) Catecholamine concentrations in plasma and organs of fetal guinea pigs during normoxaemia, hypoxaemia and asphyxia. J Dev Physiol 15:145–152

Cardiovascular Effects of Endotoxin and Asphyxia in Fetal Sheep Near Term

A. Jensen, U. Lang, and G. Braems

A prospective sonographic study on the incidence of cerebral hemorrhage in 2781 babies born in our unit between 1984 and 1986 revealed that maternal signs of infection during parturition (temperature >38 °C) are associated with a high risk for the newborns to develop peri- or intraventricular hemorrhages (Jensen et al. 1992). Since most of these infections are caused by gram-negative bacteria that produce endotoxin, we wished to determine whether endotoxin affects both blood flow and oxygen delivery to the fetal brain and whether it interferes with circulatory centralization during asphyxia.

Methods

To answer these questions, five fetal sheep were chronically catheterized at 130 ± 3.5 days of gestation (term is at 147 days) as described previously (Jensen and Lang, 1988). The fetuses were studied 3 days after the operation. After control measurements of both blood flow distribution (microsphere method. 15 μ diameter) and blood gases, acid-base balance, hemoglobin, catecholamines, vasopressin, angiotensin II, lactate and glucose concentrations, and white blood cell count, an injection of 55 ± 7 μg endotoxin (LPS) was given into the fetal inferior vena cava. Thirty minutes later, asphyxia was induced by occlusion of the caudal maternal aorta for 2 min. The measurements were repeated after 0, 2, 4, 30, and 60 min. Maternal and fetal arterial blood pressure, fetal heart rate, and amniotic fluid pressure were recorded continuously throughout the experiment.

Results

After injection of endotoxin there was an increase in arterial blood presure, and of concentrations of hemoglobin, glucose, lactate, catecholamines, vasopressin, and angiotensin II and a decrease in base excess and granulocyte counts, while PO_2 did not change. Blood flows to a number of fetal organs decreased, including those to the brain (Fig. 1), placenta, carcass, skin, and scalp; only those to the lung (massive), heart, pituitary, gastro-

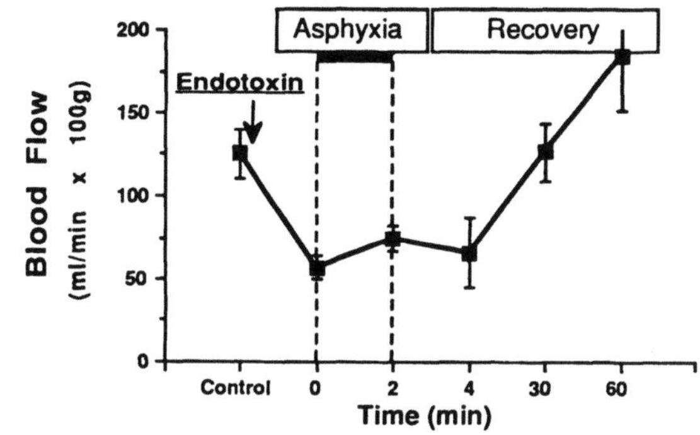

Fig. 1. Changes in blood flow to the brain after endotoxin administration and acute asphyxia in five fetal sheep near term

intestinal tract, pancreas, and liver increased. During asphyxia, the blood flow to the brain did not increase and that to the peripheral organs decreased further, including that to the gastrointestinal tract. These changes recovered partially within 1 h, including those in the lung.

Conclusions

We conclude that endotoxemia causes vasoconstriction in the brain, carcass, skin and umbilical circulation, and a massive vasodilatation in both lungs and pituitary, whereas vascular resistance in the gastrointestinal tract is not changed. We also conclude that circulatory centralization during asphyxia is blunted by endotoxin. Thus, among other causes, reduced cerebral oxygen delivery during endotoxemia may explain in part why cerebral hemorrhage is frequent in babies born with signs of infection.

References

Jensen A, Klingmüller V, Künzel W, Sefkow S (1992) Hirnblutungsrisiko bei Früh- und Reifgeborenen: Geburtshilfe und Frauenheilkunde 52: 6–20

Jensen A, Lang U (1988) Dynamics of circulatory centralization and release of vasoactive hormones during acute asphyxia in intact and chemically sympathectomized fetal sheep. In: Künzel W, Jensen A (eds) The endocrine control of the fetus. Physiologic and pathophysiologic aspects. Springer, Berlin, Heidelberg, New York, pp 135–149

III Regulatory Mechanisms of Fetal Oxygen Consumption

Indications for the Involvement of a Hemoprotein as an Oxygen Pressure Sensor Protein in the Carotid Body

H. ACKER

Introduction

The carotid body located at the carotid sinus can transduce changes in oxygen pressure (pO_2) in the arterial blood into nervous signals regulating respiration and circulation in order to avoid hypoxic situations in the body. The mechanism of the transducing process is still a matter of discussion. A generally accepted concept defines this process as a pO_2-dependent transmitter release from carotid body type I cells which generates action potentials in postsynaptic nerve endings of the sinus nerve (for review see Acker 1989).

Recent voltage clamp studies have shown that outward K^+ currents of type I cells are inhibited by hypoxia (Hescheler et al. 1989; Lopez-Lopez et al. 1989), and this was substantiated by the observation that the open probability of potassium channels in these cells decreased under hypoxia (Delpiano and Hescheler 1989). Na^+ and Ca^{++} currents are unaffected by hypoxia (Hescheler et al. 1989; Lopez-Lopez et al. 1989). The pO_2 dependence of K^+ channels explains the depolarization of type I cells under hypoxia, which might lead, by opening of voltage-dependent calcium channels, to an influx of calcium into the cytosol of these cells. The resulting increased intracellular calcium level would facilitate transmitter release under participation of intracellular calcium stores (Pietruschka 1985; Biscoe et al. 1989; Delpiano and Acker 1989; Shaw et al. 1989). The molecular mechanism of this pO_2 monitoring process is not known but several hypotheses have been developed.

In a metabolic hypothesis, the respiratory chain of the carotid body mitochondria has often been considered to be the most probable candidate for a pO_2-dependent transmitter release. Anichkow and Belen'kii (1963) as well as Joels and Neil (1963) assumed that chemosensory excitation, especially under hypoxia is caused by a decrease in adenosine triphosphate (ATP) levels in the carotid body tissue. Biscoe (1971) proposed that energy depletion under hypoxia triggers the nerve discharge by producing membrane instability in the sensory nerve endings. Using several inhibitors and uncouplers of the respiratory chain, Mulligan et al. (1981) gave further support to the idea that oxidative phosphorylation is involved in the carotid body chemoreception. However, Obeso et al. (1989) showed that carotid body excitation by cyanide mimicking the hypoxic chemoreceptor response does not alter the ATP content of the organ, thereby contradicting the metabolic hypothesis. This contradiction was supported by Acker and Starlinger (1984) showing that the ATP content of the carotid body does not change under hypoxia. Mills and Jöbsis (1972) discovered from their photometric studies on the respiratory chain of the carotid body with the dual wavelength method an unusually low affinity component for oxygen of cytochrome *aa*$_3$, which might act as a pO_2 sensor. Acker and Eyzaguirre (1987) repeated these studies with a whole spectral analysis in the visible range of the mouse carotid body showing unknown hypoxia-induced light transmission changes, which might have contaminated the dual wavelength measurements. The present paper describes the same photometric measurements on superfused mouse and rat carotid bodies with the indication that these hypoxia-induced light transmission changes might be caused by an oxidase similar to the respiratory burst oxidase as described in neutrophiles (Cross and Jones 1986).

Materials and Methods

After prolonged flushing of the common carotid arteries with Macrodex 6 % (Knoll, FRG) to eliminate red cells from the tissue, the carotid bodies and their immediate vessels were excised from mice and rats anesthetized with pentobarbital and heparinized with Liquemin. The carotid bodies were denuded of all other structures and placed in a small lucite chamber mounted on the stage of an upright photomicroscope (Zeiss, FRG). The cleaned organs were then superfused with modified Locke's solution, as described by Delpiano and Acker (1989), equilibrated with different O_2 mixtures, and viewed with a 40 × water immersion lens and a 10 × ocular lens. Oxygenation of the superfusion medium was controlled close to the tissue with a needle pO_2 electrode (Acker and Eyzaguirre 1987), the tip of which was located near the lens. Temperature was routinely maintained at 34°–35 °C, and pH was kept at 7.36.

For light absorption measurements, light from a halogen lamp (12 V, 100 W), filtered by a motor-driven monochromator (M20, Zeiss, FRG) transilluminated the organ continuously between wavelengths of 410 and 620 nm with a half bandwidth of 5–8 nm. Light signals crossing the preparation were recorded by a photomultiplier (Knott, FRG) placed on the third ocular tube of the microscope's trinocular head. The photomultiplier signals together with the wavelength information from the monochromator were stored in a computer (Perkin Elmer 7/32). Light intensity changes of $\times 10^{-3}$ optical density (0.12 ± 0.011) can be detected with this device. To demonstrate the spectral characteristics of the carotid body, the organ maintained under well oxygenated conditions served as a baseline from which spectra under hypoxic conditions could be automatically substracted by the computer. Only 1 min was required to record a whole spectrum. For each single experimental situation in one carotid body, five spectra were recorded and averaged, counting as one experiment.

Results and Discussion

Figure 1 shows hypoxic difference spectra (hypoxia) as a mean ± SE of 25 rat carotid bodies and 5 mouse carotid bodies. The absorbance spectrum during

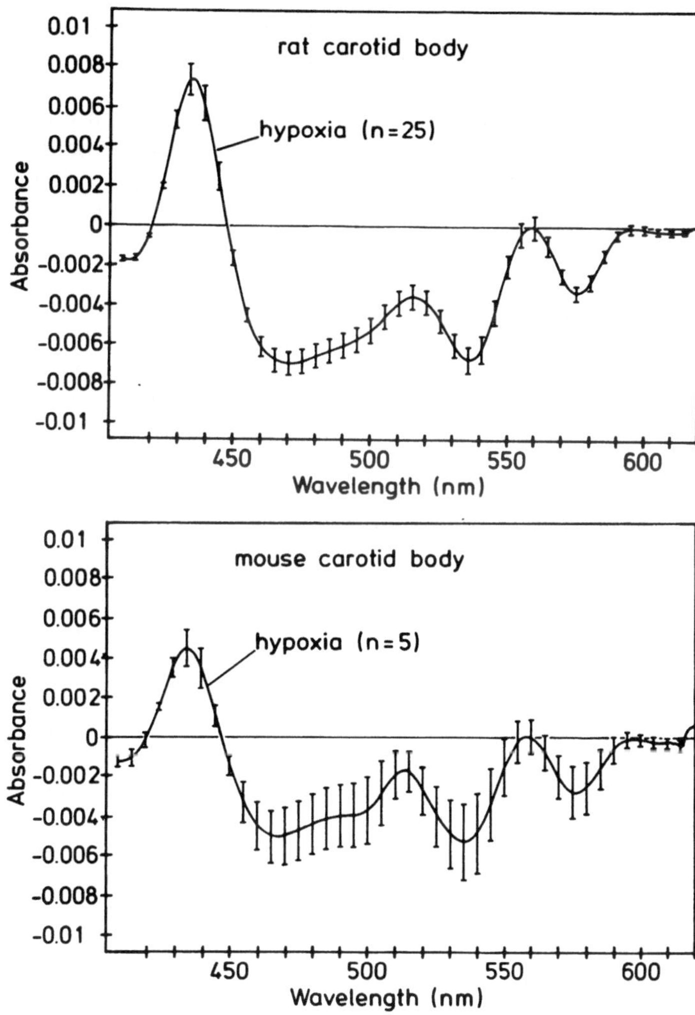

Fig. 1. Oxidized minus reduced difference spectra of the rat (*upper trace*) and mouse (*lower trace*) carotid body with the wavelength (nm) on the x-axis and the absorbance change on the y-axis

superfusion with a pO_2 of 340 torr served as a baseline in these experiments from which spectra with hypoxic superfusion (pO_2, 20 torr) were utomatically subtracted. The spectra show a typical absorbance maximum at 436 nm and a shoulder at 559 nm, with a bleaching of the absorbance peaking at 413, 470, 535, and 575 nm.

Since the form of the spectrum drew attention to the described hypoxic difference absorbance spectrum of NADPH oxidase in neutrophils (Cross and Jones 1986), Acker et al. (1989) carried out the same experiments with diphenyliodonium (DPI; 10 μM), which is known to be a special inhibitor of a 45 kDa protein regulating the activity of NADPH oxidase. NADPH oxidase consists of two redox couples, FAD/FADH and cytochrome *b* reduced/oxidized, which carry electrons from NADPH to oxygen, generating superoxide anions (O_2^-). Since DPI inhibited the hypoxic difference spectrum in the rat carotid body (Acker et al. 1989), it was suggested that a hemeprotein such as NADPH oxidase is located in the carotid body. In a more detailed study, Cross et al. (1990) could show that the light absorbance spectra of the rat carotid body suggest the presence of a *b*-type cytochrome, which was confirmed by pyridine hemochrome and CO spectra. The quantity of cytochrome *b* was estimated to be 310 pmol mg^{-1} protein. This hemoprotein is capable of H_2O_2 formation which can be inhibited by DPI. The hypoxia-induced nervous chemoreceptor discharge increase and the reduction of FAD and NADP was also inhibited by DPI.

These results suggest that a hemeprotein acts as a pO_2 sensor protein in the carotid body probably inducing the pO_2 chemoreceptor process by H_2O_2 formation. However, future experiments have to give stronger support for this speculative pO_2-dependent H_2O_2 formation and its possible importance for the pO_2 chemoreception in the carotid body. It would be interesting to study the following two hypotheses in this context:

1. The involvement of H_2O_2 in pO_2 chemoreception was proposed by Sies (1977) with the model that glutathione peroxidase degrades hydroperoxides and thereby induces a transition in the 2GSH/GSSG redox system, which as the major cellular pool of mobile sulfhydryl groups can change protein formations or influence phosphorylase and ATPase activity with the consequence of, for instance, altered cell membrane channel conductivity or function of intracellular calcium stores.
2. Another possibility for the regulation of ion channel conductivity is that the hemeprotein in the carotid body is linked to guanylate cyclase, as described in the bovine lung (Gerzer et al. 1981), regulating the cGMP content, which is known to decrease in the carotid body under hypoxia (Wary et al. 1989).

Understanding of the pO_2 chemoreceptor process in the carotid body and the characterization of the responsible hemeprotein for this process is important to elucidate similar chemoreceptor processes in other cells of the body.

References

Acker H (1989) Am Rev Physiol 51:835–844

Acker H, Eyzaguirre C (1987) Brain Res 409:380–385

Acker H, Starlinger H (1984) Neurosci Lett 50:175–179

Acker H, Dufau E, Huber J, Sylvester D (1989) FEBS Lett 256:75–78

Anichkow SW, Belen'kii MC (1963) In: Pharmacology of carotid body chemoreceptors. Pergamon, Oxford, pp 187–192

Biscoe T (1971) Physiol Rev 51:437–495

Biscoe TJ, Duchen MR, Eisner DA, O'Neill SC, Valdeomillos M (1989) Physiol (Lond) 416:421–434

Cross AR, Jones OTJ (1986) Biochem J 237:111–116

Cross AR, Henderson L, Jones OTJ, Delpiano MA, Hentschel J, Acker H (1990) Biochem J 272:743–747

Delpiano MA, Acker H (1989) Brain Res 482:235–246

Delpiano MA, Hescheler J (1989) FEBS Lett 249:195–198

Gerzer R, Böhme E, Hofmann F, Schultz G (1981) FEBS Lett 132:71–74

Hescheler J, Delpiano MA, Acker H, Pietruschka F (1989) Brain Res 486:79–88

Joels N, Neil E (1963) Br Med Bull 19:21–24

Lopez-Lopez J, Gonzales C, Urena J, Lopez-Barneo J (1989) J Gen Physiol 93:1001–1015

Mills E, Jöbsis FF (1972) J Neurophysiol 35:405–428

Mulligan E, Lahiri S, Storey BT (1981) J Appl Physiol 51:438–446

Obeso A, Almaraz L, Gonzales C (1989) Brain Res 481:250–257

Pietruschka F (1985) Brain Res 347:140–143

Shaw K, Montague W, Pallot DJ (1989) Biochem Biophys Acta 1013:42–46

Sies H (1977) In: Tissue hypoxia and ischemia. Plenum, New York, pp 51–66

Wary WJ, Cheng GF, Dinger BG, Fidone SJ (1989) Neurosci Lett 105:164–168

Role of Erythropoietin in the Regulation of Red Cell Production in Children Before and After Birth

K.-U. Eckardt, A. Kurtz, and C. Bauer

An increasing number of hormones and cytokines are currently being recognized to influence maturation and differentiation of erythroid progenitor cells. Although their relative roles and complex interactions are not yet understood, conclusive evidence indicates that during adulthood the glycoprotein hormone erythropoietin (EPO) is the major regulator of red cell formation (for review, see Eckardt and Bauer 1989). In the healthy adult, basal EPO levels of about 10–25 mU/ml are essential to maintain erythropoiesis at a rate that compensates for daily red cell degeneration. Furthermore, increased EPO production under anemic or hypoxic hypoxia stimulates erythropoiesis in an attempt to either restore red cell mass to normal or to induce compensatory polycythemia. In chronic anemias, this oxygen-dependent EPO formation is reflected by an inverse exponential relationship between hematocrit and serum EPO levels (Cotes 1982; Rege et al. 1982; Garcia et al. 1982; Erslev et al. 1987). The kidneys are the main production sites for EPO during adult life, and renal disease is generally accompanied by an anemia due to inappropriately low EPO levels. In addition, some EPO is produced extrarenally, probably in the liver, and maintains a detectable EPO level in serum after bilateral nephrectomy. A complete EPO deficiency has only been observed in animals immunized against EPO and was found to result in lethal anemia.

During childhood, erythropoiesis not only subsitutes for red cell destruction but, additionally, the erythron has to be adjusted to the growth of the organism. Furthermore, red cell volume undergoes characteristic changes in relation to blood volume during development, resulting in age-dependent variations of hemoglobin concentrations. The main alterations occur during three stages: fetal life, early postnatal life, and the period of accelerated growth during puberty. The relative roles of EPO and other hormones in the regulation of erythropoiesis during these periods is incompletely understood. Studies on EPO formation were limited as long as only bioassays for the hormone were available, since these require large sample volumes and have a detection limit which generally lies above normal serum values. However, in recent years the development of sensitive immunoassays has allowed the determination of EPO levels in healthy and diseased children at various ages and thereby provided further insight into the formation of this hormone during childhood.

Fetal Life

Immunoreactive EPO has been demonstrated in human fetal blood as early as at 19 weeks gestation (Thomas et al. 1983) EPO is also detectable in amniotic fluid, and a close correlation was found between EPO concentrations in amniotic fluid and umbilical plasma (Teramo et al. 1987). At mid pregnancy, cord serum EPO levels are below those of children older than 3 months or of adults. Thereafter, between 20 weeks and term, an about fourfold increase in EPO occurs, and at term EPO levels in cord blood are higher than during later life (Thomas et al. 1983) (Fig. 1). This increase in EPO in the fetus is accompanied by a continuous increase in hemoglobin concentrations (for review, see Finne and Halvorsen 1972; Thomas et al. 1983). Although this finding cannot prove any causality, it may suggest that EPO plays in important role in the regulation of fetal red cell production.

The origin of EPO in the fetal circulation is not completely clear. Variable, up to tenfold concentration differences between human fetal and maternal blood in either direction (Thomas et al. 1983, and our own unpublished

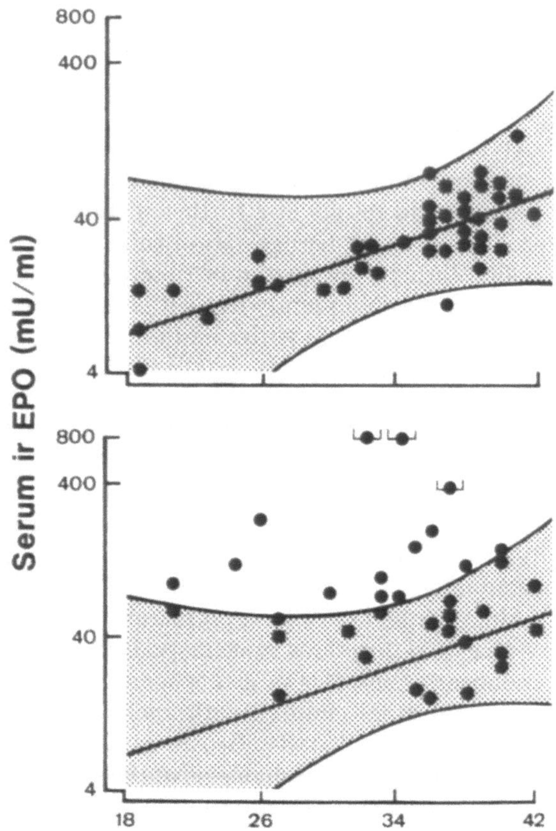

Fig. 1. Estimates of immunoreactive cord serum erythropoietin (EPO) and gestation in normal (*upper*) and abnormal (*lower*) pregnancy (adapted from Thomas et al. 1983). Abnormal pregnancies included twins, maternal complications such as hypertension, pre-eclampsia, intrauterine growth retardation, diabetes, acute fetal distress in labor, fetal abnormalities, and rhesus hemolytic disease. *Shaded area* represents 95 % confidence band for a single observation of EPO in normal pregnancy. Particularly high EPO levels were found in two infants with intrauterine growth retardation (>800 and 150 mU/ml), in an infant with Down's syndrome (>400 mU/ml), and in infants with severe rhesus incompatibility before transfusion (>800, 340 and 118 mU/ml)

observations) indicate that EPO does not equilibrate across the placenta. Those cases in which fetal EPO levels are several times higher than maternal levels, suggest an autonomous EPO production in the fetus. In mice, however, some maternofetal transfer of EPO has been found (Koury et al. 1988), and placental receptor sites for EPO have been demonstrated (Sawyer et al. 1989). For the human placenta, no direct data on the transfer of EPO are yet available.

Irrespective of the possibility that some EPO in the fetal circulation may be of maternal origin, several studies indicate that the fetus can respond to hypoxia with an increase in EPO formation (Fig. 1). Thus, elevated cord serum EPO levels were determined in the presence of fetal hemolytic anemia due to rhesus incompatibility (Thomas et al. 1983; Voutilainen et al. 1989) or placental insufficiency (Thomas et al. 1983) and cord serum EPO at birth was found to be correlated with signs of fetal distress (umbilical artery pH or apgar score) (Voutilainen et al. 1989; Eckardt et al. 1990). Although it has not been precisely assessed to which extent fetal erythropoiesis may be increased by elevated EPO levels, pathological pregnancies may be associated with fetal polycythemia (for review, see Finne and Halvorsen 1972), suggesting that the feedback system controlling blood oxygen content via oxygen-dependent EPO formation is already effective in the fetus.

Early Postnatal Life

At birth, blood is shifted from placental circulation into the fetus. A subsequent decrease in plasma volume in the newborn restores total blood volume and results in a further increase of the hemoglobin concentrations achieved during late gestation (normal mean value of newborns, about 20 g/dl). Within a few days of birth, erythropoiesis decreases to subnormal adult levels and subsequently leads to a fall in hemoglobin concentrations during the following weeks with a nadir in the 3rd month (normal mean hemoglobin concentration, about 11.5 g/dl) (for review, see Finne and Halvorsen 1972). This decrease in erythropoiesis is accompanied by a reduction in serum EPO levels, and EPO levels during the first 2 months appear to be lower than in any other postnatal period (Hellebostad et al. 1988; Eckardt et al. 1990) (Fig. 2). Preterm infants experience an even greater fall in hemoglobin concentration during the first weeks than term infants, with values frequently as low as 7–8 g/dl at 1–3 months of age. During this anemia of prematurity, EPO levels are generally also below those during later life (Brown et al. 1983, 1984). Although EPO levels in premature infants vary inversely with hemoglobin values and thus show some responsiveness to changes in oxygen supply, EPO concentrations at any given hemoglobin level are remarkably lower than those of older children with anemias resulting from bone marrow failure (Stockman et al. 1984; Brown et al. 1984).

Regarding the mechanisms of the early decline in EPO after birth and the decreased EPO response to reductions in blood oxygen content, it has to be

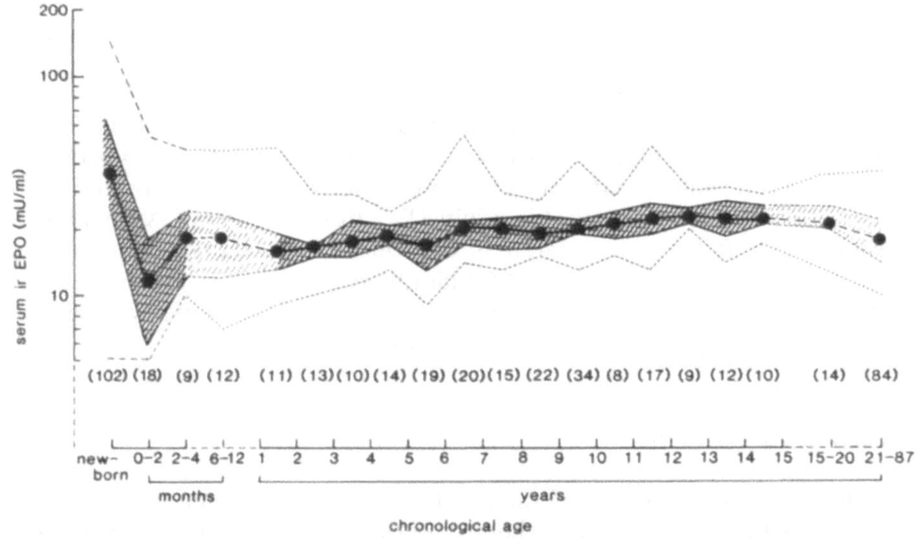

Fig. 2. Estimates of serum immunoreactive erythropoietin (EPO) in healthy children in relation to age (from Eckardt et al. 1990). For comparison, values of healthy adolescents and healthy adults are included. Values are geometric mean, interquartile range, and total range; *numbers in parentheses* are numbers of observations

considered that the increase in the proportion of adult hemoglobin, from 20 % to 30 % at birth to 97 % at the age of 5 months, results in a decreased oxygen affinity of red cells. This facilitates unloading of oxygen, and the decline in EPO levels may in part be attributable to this improvement in oxygen delivery. However, this explanation does not account for the fact that EPO and hemoglobin concentrations after birth fall to values lower than those in adults. The observation that EPO levels do not increase, as they would during later life in the presence of similarily low hemoglobin concentrations, indicates, rather, that the sensitivity of EPO formation towards changes in oxygen supply is significantly diminished during this early postnatal period. This assumption is further supported by the observation that in infants with congenital anemias EPO levels decline temporarily during the first weeks after birth, despite their oxygen transport capacity being continuously impaired (unpublished data).

It is tempting to speculate that the temporary insensitivity of EPO formation towards reductions in hemoglobin concentrations in normal children and, even more pronounced, in preterm infants might coincide with the shift in the production site of EPO, which is thought to occur around birth. The liver is considered the main production site during fetal life, and the EPO gene has in fact been cloned from human fetal liver (Jacobs et al. 1985). However, the mechanisms of the transition in EPO production from liver to kidney are unknown, and, furthermore, in humans no data are available on the relative contribution of liver and kidney derived EPO at different stages of development.

After the age of 3 months, serum EPO values in healthy children are essentially constant and the same as in adults (Hellebostad et al. 1988; Eckardt et al. 1990) (Fig. 2). Furthermore, the inverse relationship between arterial oxygen content and EPO levels in children older than 3 months is not significantly different from that in adults (Eckardt et al. 1990). As in adults, chronic renal failure in children is associated with an anemia due to inappropriately low EPO levels (Aikhionbare et al. 1987; Eckardt et al. 1990), providing indirect evidence that at this stage the kidneys have become the predominant production sites of the hormone.

Accelerated Growth

During the growth period of mammals, red cell mass increases in strict proportion to the body mass, thereby ensuring that an increased oxygen consumption due to growth processes is matched by increased oxygen transport capacity (Garcia 1957). This proportionality becomes most important when growth accelerates, i.e., during puberty. EPO serum values in healthy children during puberty do not change (Hellebostad et al. 1988; Eckardt et al. 1990) (Fig. 2), and in rats even a slight decline in EPO values was observed during accelerated growth (Kurtz et al. 1990). This finding suggests that EPO is not the only determinant governing the expansion of red cell mass during this period of postnatal development. Furthermore, experimentally induced polycythemia in rodents during rapid growth was found to suppress EPO levels (Sanengen et al. 1989) but, in contrast to the situation in adult animals, does not blunt erythropoiesis (Sanengen and Halvorsen 1985; Sanengen et al. 1987) and erythropoietic bioactivity in serum (Sanengen et al. 1989). Among factors other than EPO, insulin-like growth factor I (IGF I), which is considered to mediate the growth-promoting activities of growth hormone (Zapf et al. 1984), appears to play a major role in the increase of erythropoiesis during growth. IGF I has been shown to stimulate erythroid progenitors in vivo (Phillipps et al. 1988) as well as in vitro (Kurtz et al. 1982; Sawada et al. 1989), and its application to hypophysectomized, growth-arrested rats caused a proportional increase of growth and erythropoiesis (Kurtz et al. 1988). Furthermore, serum levels of IGF I rise during puberty, and in rats a positive linear correlation was found between serum IGF I levels and erythropoiesis, as assessed by incorporation of radiolabeled iron into newly formed red cells (Kurtz et al. 1990) (Fig. 3).

In addition to growth-related alterations, a sex-related difference in erythropoiesis also develops during puberty. There is no evidence that the higher red cell mass in males after puberty is due to increased EPO stimulation. Two studies found no difference in EPO levels during puberty in boys and girls (Hellebostad et al. 1988; Eckardt et al. 1990), and in adults

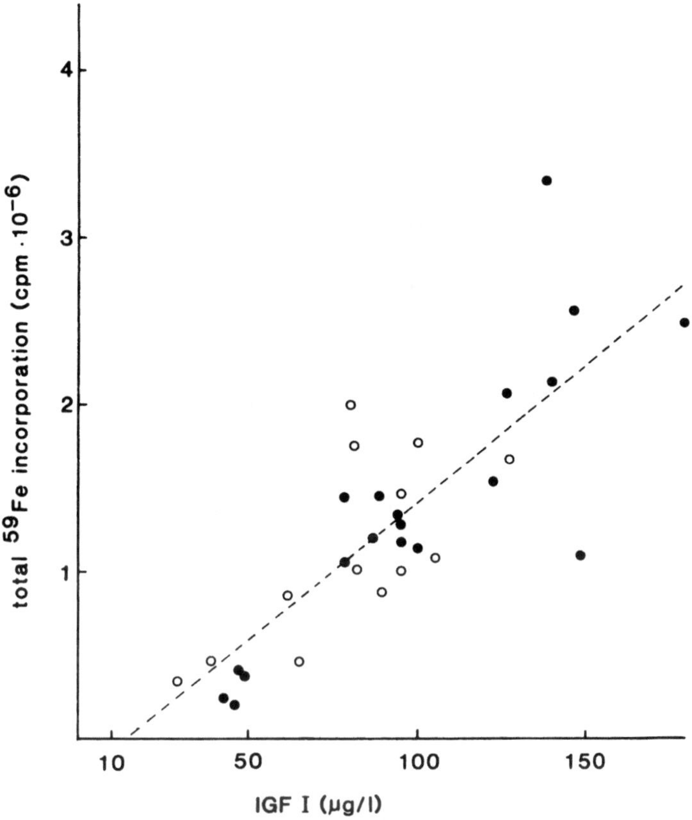

Fig. 3. Relationship of total incorporation of ^{59}Fe into newly formed red blood cells (reflecting erythropoiesis) to serum insulin-like growth factor (IGF) I levels in male (●) and female (○) rats during accelerated growth) (from Kurtz et al. 1990). Values fit the linear regression: total ^{59}Fe-inc. (cpm 10^{-6}) = $-0.266 + 0.012 \times$ [IGF-I] (μg/l); $r = 0.81$; $p < 0.0001$

some investigators also found no sex-related difference (Cotes 1982; Rege et al. 1982), while others reported even slightly lower EPO levels in males (Garcia et al. 1982). Since androgens have been shown to directly stimulate erythropoiesis (Adamson 1977), it appears possible that their effect accounts for the higher hemoglobin concentrations in males.

Clinical Implications

Since normal EPO values in childhood have recently been established, the determination of serum EPO concentrations in children may contribute to the diagnosis and the understanding of hematopoietic disorders. Furthermore, it appears conceivable that EPO determinations in fetal serum and/or amniotic fluid may proof valuable in assessing fetal hypoxia and monitoring pregnancies at risk. The advantage of EPO determinations in this situation may be that they reflect an average alteration in fetal oxygen supply rather than acute, transient changes.

The finding of very low EPO levels in premature infants provides a rational basis for an attempt to treat the anemia of prematurity with recombinant human EPO. Clinical studies are currently underway to assess the benefit of EPO application in this situation, and their outcome will also allow conclusions on the pathogenicity of this anemia.

Note added in proof: Since preparation of this manuscript several investigators reported first promising results on the treatment of premature infants with recombinant human Epo (Halpérin DS, Wacker P, Lacourt G, Félix M, Babel J-F, Aapro M, Wyss M (1990)). Effects of recombinant human erythropoietin in infants with the anemia of prematurity: a pilot study. *J Pediatr 116:* 779–786; Shannon KM, Mentzer WC, Abels RI, Freeman P, Newton N, Thompson D, Sniderman S, Baleard R, Phibbs RH (1991) Recombinant human erythropoietin in the anemia of prematurity: results of a placebo-controlled pilot study. J Pediatr 118: 949–955; Ohls RU, Christensen RD (1991) A randomized trial of erythropoietin administration versus erythrocyte transfusion in patients with the anemia of prematurity (Abstr.) Clin Res 39: 12A.

References

Adamson JW (1977) Steroid metabolites and hematopoiesis. In: Fisher JW (ed) Kidney hormones, vol 2: erythropoietin. Academic, London, pp 437–462

Aikhionbare HA, Winterborne MW, Gyde OH (1987) Erythropoietin in children with chronic renal failure on dialytic and non-dialytic therapy. Int J Pediatr Nephrol 8:9–14

Brown MS, Phibbs RH, Garcia JF, Dallman PR (1983) Postnatal changes in erythropoietin levels in untransfused premature infants. J Pediatr 103:612–617

Brown MS, Garcia JF, Phibbs RH, Dallman PR (1984) Decreased response of plasma immunoreactive erythropoietin to "available oxygen" in anemia of prematurity. J Pediatr 105:793–798

Cotes PM (1982) Immunoreactive erythropoietin in serum. I. Evidence for the validity of the assay method and the physiological relevance of estimates. Br J Haematol 50:427–438

Eckardt K-U, Bauer C (1989) Erythropoietin in health and disease. Eur J Clin Invest 19:117–127

Eckardt K-U, Hartmann W, Vetter U, Pohlandt F, Burghardt R, Kurtz A (1990) Serum immunoreactive erythropoietin of children in health and disease. Eur J Pediatr 199:450–464

Erslev AJ, Wilson J, Caro J (1987) Erythropoietin titers in anemic, nonuremic patients. J Lab Clin Med 109:429–433

Finne PH, Halvorsen S (1972) Regulation of erythropoiesis in the fetus and newborn. Arch Dis Child 47:683–687

Garcia JF (1957) Changes in blood, plasma and red cell volume in the male rat, as a function of age. Am J Physiol 190:19–24

Garcia JF, Ebbe SN, Hollander L, Cutting HO, Miller ME, Cronkite EP (1982) Radioimmunoassay of erythropoietin: circulating levels in normal and polycythemic human beings. J Lab Clin Med 99:624–635

Hellebostad M, Haga P, Cotes M (1988) Serum immunoreactive erythropoietin in healthy normal children. Br J Haematol 70:247–250
Jacobs K, Shoemaker C, Rudersdorf R, Neill EF, Kaufman RJ, Mufson A, Seehra J, Jones SS, Hewick R, Fritsch EF, Kawakita M, Shimaza T, Miyake T (1985) Isolation and characterization of genomic and cDNA clones of human erythropoietin. Nature 3143:806–810
Koury MJ, Bondourant MC, Graber SE, Sawyer ST (1988) Erythropoietin messenger RNA levels in developing mice and transfer of ^{125}I-erythropoietin by the placenta. J Clin Invest 82:154–159
Kurtz A, Jelkmann W, Bauer C (1982) A new candidate for the regulation of erythropoiesis. Insulin-like growth factor I. FEBS Lett 149:105–108
Kurtz A, Zapf J, Eckardt K-U, Clemons G, Froesch ER, Bauer C (1988) Insulin-like growth factor I stimulates erythropoiesis in hypophysectomized rats. Proc Natl Acad Sci USA 85:7825–7829
Kurtz A, Matter R, Eckardt K-U, Zapf J (1990) Erythropoiesis, serum erythropoietin and serum IGF I in rats during accelerated growth. Acta Endocrinol (Copenh) 122:323–328
Philipps AF, Persson B, Hall K, Lake M, Skottner A, Sanengen T, Sara V (1988) The effects of biosynthetic insulin-like growth factor-I supplementation on somatic growth, maturation and erythropoiesis in the neonatal rat. Pediatr Res 23:298–305
Rege AB, Brookins J, Fisher JW (1982) A radioimmunoassay for erythropoietin: serum levels in normal human subjects and patients with hemopoietic disorders. J Lab Clin Med 100:829–843
Sanengen T, Halvorsen S (1985) Regulation of erythropoiesis during rapid growth. Br J Haematol 61:273–279
Sanengen T, Myhre K, Halvorsen S (1987) Erythropoietic factors in plasma from neonatal mice. In vivo studies by the exhypoxic polycythemic mice assay for erythropoietin. Acta Physiol Scand 120:381–386
Sanengen T, Clemons GK, Halvorsen S, Widness JA (1989) Immunoreactive erythropoietin and erythropoiesis stimulating factor(s) in plasma from hypertransfused neonatal and adult mice. Studies with a radioimmunoassay and a cell culture assay for erythropoietin. Acta Physiol Scand 135:11–16
Sawada K, Krantz SB, Dessypris EN, Koury ST, Sawyer ST (1989) Human colony-forming units-erythroid do not require accessory cells, but do require direct interaction with insulin-like growth factor I and/or insulin for erythroid development. J Clin Invest 83:1701–1709
Sawyer ST, Sawada K, Krantz SB (1989) Receptors for erythropoietin in mouse and human erythroid cells and placenta. Blood 74: 103–109
Stockman JA III, Graeber JE, Clark DA, McClellan K, Garcia JF, Kavey REW (1984) Anemia of prematurity: determinants of the erythropoietin response. J Pediatr 105:786–792
Teramo KA, Widness JA, Clemons GK, Voutilainen PEJ, McKinlay S, Schwartz R (1987) Amniotic fluid erythropoietin correlates with umbilical plasma erythropoietin in normal and abnormal pregnancy. Obstet Gynecol 69:710
Thomas RM, Canning CE, Cotes PM, Linch DC, Rodeck CH, Rossiter CE, Huehns ER (1983) Erythropoietin and cord blood haemoglobin in the regulation of human fetal erythropoiesis. Br J Obstet Gynaecol 90:795–800
Voutilainen PEJ, Widness JA, Clemons GK, Schwartz R, Teramo KA (1989) Amniotic fluid erythropoietin predicts fetal distress in Rh-immunized pregnancies. Am J Obstet Gynecol 160:429–434
Zapf J, Schmid C, Froesch ER (1984) Biological and immunological properties of insulin-like growth factors (IGF) I and II. Clin Endocrinol Metab 13:3–30

Oxygen Availability Determines Oxygen Consumption of Fetal Skeletal Muscle Cells in Monolayer Culture: Preliminary Report

G. Braems and A. Jensen

Introduction

In a previous study on the effects of asphyxia on unanesthetized fetal sheep Jensen et al. (1987) showed that reduced oxygen delivery to peripheral organs reduces total oxygen consumption. This observation suggested that oxygen itself may be a determinant of fetal oxygen consumption. To test this hypothesis, we developed an in vitro model, which enabled us to measure oxygen consumption of fetal cells in monolayer culture during normoxia and during various degrees of hypoxia.

Material and Methods

Cell Culture

Skeletal muscle specimens were obtained from guinea pig fetuses (Purbright white) at 60 ± 2 days (term is at 68 days) under sterile conditions. These were rinsed with Hank's balanced salt solution (HBSS), cut into pieces, and treated with trypsin to dissociate fetal skeletal muscle cells. After centrifugation, the cells were resuspended in a medium containing basal medium (Eagle) with Hank's salts (BME), 10 % fetal calf serum, 1 % antibiotics, 1 % glutamine, and a buffer (25 mM HEPES). The cultures were maintained at 37 °C. The medium was exchanged every 3 or 4 days. Depending on the density of the cell population, cells were trypsinated and transferred to new flasks. The mean number of transfers was 4.2 ± 2 (SD). All materials from Gibco Europe (Karlsruhe, FRG).

Experimental Protocol

At least 24 h before the experiment, the cells were transferred to Petri dishes where they adhered and grew as a monolayer. The Petri dish was converted into a specially designed air-tight chamber using a rubber ring and a metal plate with inlet and outlet manifolds, which allowed the salt solution to perfuse or to bypass the cell culture (Peterson and Walum 1985). The air-tight chamber was placed in a warm waterbath maintained at 37 °C using preincubated HBSS as perfusate. The oxygen tension of the perfusate was measured continuously by a polarographic electrode (Transoxode, Dräger and Oxymonitor, Hellige, FRG) connected to a chart recorder. Samples were taken from the perfusate to measure pH and PCO_2 in a blood gas analyzer (Model 178, Corning Europe, FRG). The polarographic electrode was calibrated before (at 0 and 150 mmHg PO_2), during (150 mmHg PO_2), and after (at 0 and 150 mmHg PO_2) the experiment. The flow rate of the perfusate was 1 ml/min. This resulted in a quick replacement of the dead space in the Petri dish without detaching cells. At the end of the experiment, the cells were dissociated from the Petri dish using trypsin. The number of dissociated cells were determined in a Neubauer counting chamber. Their vitality (usually >95 %) was demonstrated by adding trypan blue. Only one experiment per monolayer culture was performed.

Measurements

To determine the oxygen consumption of the cultured cells in the Petri dish, the difference in PO_2 across the Petri dish was measured under steady state conditions. The difference in PO_2 was calculated as follows: PO_2 of the perfusate (control or hypoxic) bypassing the cell culture minus PO_2 of the same perfusate perfusing the cell culture. Oxygen consumption per cell per time was calculated as follows: difference in PO_2 × flow rate × concentration of dissolved oxygen at a given measured PO_2/number of cells. The concentration of dissolved oxygen in a salt solution was calculated using both PO_2 and the solubility of oxygen (Henry's law). The solubility of oxygen amounts to 5.02 μl O_2/ml in a salt solution that is equilibrated to air at 760 mmHg atmospheric pressure and at 37 °C.

To assess the change of cellular oxygen consumption during hypoxia, two stocks of perfusate were used. The control perfusate was equilibrated with air for at least 12 h; PO_2 was approximately 150 mmHg. Hypoxic perfusates were equilibrated with various gas mixtures containing reduced oxygen tensions to achieve PO_2 between 10 and 140 mmHg. There were no significant differences in pH or PCO_2 between control and hypoxic perfusates. The change in cellular oxygen consumption during hypoxia was expressed as percent of control.

Results

In 23 experiments on 23 cell cultures, the mean oxygen consumption of the fetal skeletal muscle cells at a PO_2 of 150 mmHg was 3.5 ± 2.2 (SD) $\times 10^{-6}$ μl O_2 per cell per hour. There was a close positive correlation between the PO_2 of the perfusate and the change in the oxygen consumption (as percent of control) of the fetal skeletal muscle cells

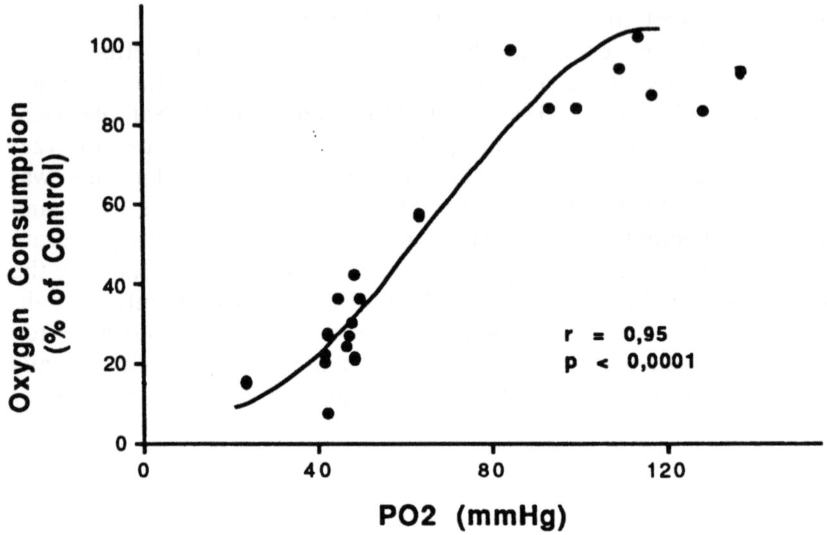

Fig. 1. Correlation between the oxygen tension (PO_2) and oxygen consumption (% of control) of fetal skeletal muscle cells in monolayer cell culture. In 23 experiments, there was a close positive correlation between the two variables. When oxygen tension was reduced there was a similar reduction in cellular oxygen consumption

($r = 0.95$; $p < 0.0001$), i.e., when PO_2 was reduced, there was a similar reduction in cellular oxygen consumption.

Comment

This is the first study to show in fetal skeletal muscle cells that oxygen is an important determinant of oxygen consumption. Furthermore, it confirms in an *in vitro* model our previous observations *in vivo* in unanesthetized fetal sheep during acute asphyxia caused by arrest of uterine blood flow (Jensen et al. 1987). During arrest of uterine blood flow, there was a rapid decrease in total oxygen consumption of the fetus, placenta, and membranes, which correlated closely with a reduction in both blood flow and oxygen delivery to peripheral fetal organs. This suggested that oxygen itself may be an important determinant of oxygen consumption. However, due to the design of the *in vivo* experiments, which did not allow us to control for various confounding factors, we were unable to determine the actual mechanism involved. Therefore, to exclude any blood-borne or circulatory influences, we developed an *in vitro* method, which allowed us to change nothing but the oxygen delivery to the cells and to measure the oxygen consumption of these cells directly.

Thus, on transition from normoxia to hypoxia or asphyxia the fetus is able to effectively reduce oxygen consumption in peripheral organs to maintain oxidative metabolism in central organs. Conversely, on transition from hypoxia to noroxia, the concomitant increase in cellular oxygen consumption and metabolic drive would warrant optimal cell function at any given state of oxygenation. This may be of particular importance for the fetal organism on transition from fetal ot postnatal live, when oxygen delivery rises.

Acknowledgements. We thank A. Peltzer, I. Dußler, and K. Valentin for their excellent technical assistance. This work was supported by the Deutsche Forschungsgemeinschaft, DFG grant Je 108/4-1.

References

Jensen A, Hohmann M, Künzel W (1987) Dynamic changes in organ blood flow and oxygen consumption during acute asphyxia in fetal sheep. J Dev Physiol 9:543–559
Peterson A, Walum E (1985) Use of a perfusion technique for measurements of respiratory activity in cultured cells. In Vitro 21:622–626

*Endothelial Cells as Part of a Vascular Oxygen-Sensing System: Hypoxia-Induced Release of Autacoids**

U. Pohl

Introduction

A continuous oxygen supply is of vital importance for the integrity and the function of cells. It is, therefore, not surprising that the circulatory system makes use various central and local control mechanisms which permit an adequate amount of tissue oxygen to be supplied over a wide range of oxygen demands. In general, when the oxygen consumption of organs is increased, e.g., during exercise, or when the oxygen content of the transport medium blood is reduced, local vasodilation leads to an increase in regional blood flow. This vasomotor response is often combined with an increase in cardiac output and an augmented vascular resistance in other organs which results in a redistribution of blood flow towards the hypoxic organ. This coordinated response to hypoxia reflects the function of effective mechanisms controlling tissue oxygen supply. An integral part of these control systems must be cells which directly or indirectly sense changes in the oxygen tension of the blood or within the tissue and respond by generating signals which induce central or local responses of the circulatory system.

Oxygen-Sensing Structures and their Localization

More highly developed organisms possess highly specialized cells which can sense the oxygen partial pressure (pO_2) of the transport medium blood. These cells are involved in the adaptive responses of the circulatory and ventilatory systems and in the increase in the oxygen-carrying capacity of the blood. Chemoreceptors in the carotid bodies, which are located in the bifurcation of the carotid arteries, and in the aortic arch respond to changes in oxygen tension by activating afferent nerve fibers (Eyzaguirre and Fidome 1980). This increased activity stimulates ventilation and contributes to a centralization of the blood volume towards vital organs. Likewise, the skeletal muscle contains elements which induce enhanced afferent neural activity

* This chapter was originally published in *Experientia* vol 46 (11–12), pp 1175–1179, Birkhäuser, Basel

during regional hypoxemia, resulting in augmented cardiac output and peripheral vasoconstriction in nonhypoxic organs (Longhurst and Zelis 1979). It should also be briefly mentioned that under conditions of hypoxemia, specialized, but not yet identified kidney cells release the glycoprotein erythropoietin, which stimulates the production of erythrocytes in the bone marrow (for review see Bauer and Kurtz 1989).

In parallel to these global control systems local mechanisms exist which regulate the matching of oxygen demands and oxygen supply within an individual organ. The nature of the oxygen-sensitive structures acting at the local tissue level is not completely understood. There is convincing evidence that under conditions of hypoxia parenchymal cells release increased amounts of metabolites such as adenosine, lactate, hydrogen ions, and potassium ions, which accumulate in the tissue. These metabolites induce relaxation of the vascular smooth muscle, which results in an increased blood flow in the hypoxic area (Berne and Rubio 1977). This mechanisms, however, implies that the cells must suffer from a certain degree of hypoxia in order to affect the organ blood flow. Another principle of control could be located in the arterial vessels itself, allowing a more rapid response to changes of blood oxygen tension without involvement of parenchymal cells. This presupposes that blood vessels are directly affected if the oxygen content or tension of the blood is reduced, either as result of hypoxemia or due to an augmented extraction of oxygen. There is indeed ample evidence that blood vessels react with a dilation (i.e. increase in organ blood flow) to a reduction of oxygen tension even in the absence of tissue-derived metabolites: In arterioles of the cheek pouch of the hamster, Jackson and Duling (1983) demonstrated a direct effect of oxygen on vascular diameter that was independent of flow conditions or the presence or absence of adjacent parenchymal tissue. Likewise, in canine skeletal muscle, vasodilation in response to a sudden onset od hypoxemia occurred as soon as the hypoxic blood reached the organ and while there were still no detectable changes of tissue oxygenation. Thus, it was concluded that the observed dilation resulted from a direct action of oxygen on the vasculature of the skeletal muscle (Pohl et al. 1982). These findings, in combination with numerous observations in isolated blood vessels (Coburn et al. 1979; Detar 1980) demonstrate direct effects of oxygen on elements of the vascular wall which might contribute to the local control of vascular tone. Endothelial cells which are in direct contact with the streaming blood have a number of properties which predestine them as effective functional vascular oxygen sensors.

Vasomotor Function of the Endothelium

The vascular endothelium lining the entire circulatory system is a highly specialized tissue. It is involved in the modulation of immune responses (Harlan 1987) and vascular cell growth (DiCorleto 1984) and in the regulation of the level of hemostatic (Stern et al. 1986), inflammatory (Bevilacqua and

Gimbrone 1987), and vasoactive (Pearson et al. 1980) agents in the blood. In addition, the endothelium releases autacoids which decisively affect vascular tone and platelet function. Many endothelium-derived autacoids have been described and characterized in the past few years, among them prostaglandins (PG) (Moncada and Vane 1979), platelet activating factor (Whatley et al. 1987), the vasoconstrictor peptide endothelin (Yanagisawa et al. 1988), and the potent vasodilator compound endothelium-derived relaxant factor (EDRF) (Busse et al. 1985; Furchgott 1983) which is identical with nitric oxide (Palmer et al. 1987). EDRF stimulates the soluble guanylate cyclase in vascular smooth muscle directly (Müsch et al. 1987) and induces vasodilation by cyclic guanosine monophosphate (cGMP)-dependent mechanisms. Many compounds, e.g., platelet-derived substances like serotonin and adenosine triphosphate (ATP) and peptides like bradykinin or substance P (Busse et al. 1985) elicit a release of EDRF by specific receptor-dependent pathways. The release of EDRF results in an endothelium-dependent vasodilation which often counteracts simultaneous direct constrictor effects of these compounds on vascular smooth muscle (Bassenge and Busse 1988). Most of these stimuli also induce the release of prostacyclin (PGI_2) from endothelial cells. PGI_2 elicits an increase of cellular cyclic adenosine monophosphate (cAMP) by a receptor-dependent stimulation of adenylate cyclase. It is a potent antiaggregatory substance and also induces vasodilation at appropriately high concentrations (Moncada and Vane 1979). However, not all vessels are sensitive to PGI_2 (Förstermann et al. 1989). Since endothelial autacoid release is not only stimulated by receptor-dependent stimuli, but also by a number of physical stimuli, it was hypothesized that endothelial cells may also respond specifically to changes in oxygen tension.

Hypoxia-Induced Endothelium-Dependent Vasodilation

The effects of hypoxia on the vasomotor function of endothelial cells were studied in isolated arteries of different species. The arterial segments were cannulated at both ends and suspended in an organ bath which was gassed to establish a pO_2 of more than 300 mmHg. The arterial lumen was perfused separately from the organ bath, either at a pO_2 of 130 mmHg (normoxia) or of 20–40 mmHg (hypoxia). This approach differs from others where endothelium and vascular smooth muscle are simultaneously made hypoxic, since only the endothelium is exposed to the reduced pO_2, whereas at the adventitia a pO_2 of more than 300 mmHg is maintained throughout the experiment.

It is consistent with a role of endothelial cells as vascular oxygen sensors that precontracted arterial segments with an intact endothelium responded with a vasodilation to luminal hypoxia, whereas this dilator response was abolished ater endothelial denudation (Fig. 1; Busse et al. 1984; Pohland Busse 1989). This endothelium-dependent vasodilation in response to luminal hypoxia is associated with an enhanced release of PGI_2 and of EDRF

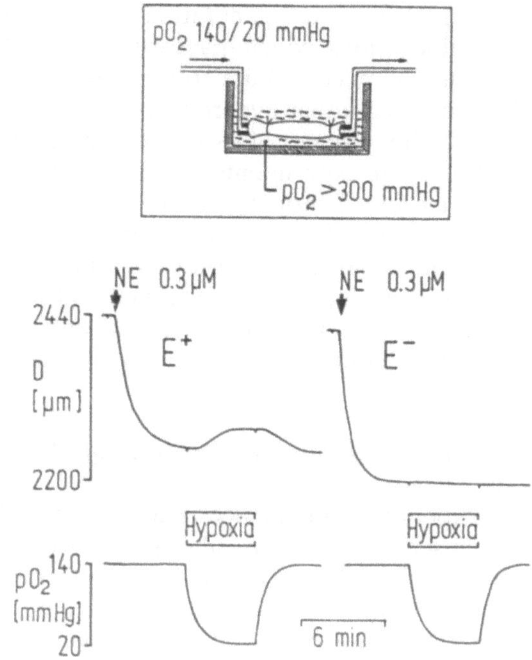

Fig. 1. Endothelium-dependent dilator response of canine femoral artery segments to selective luminal hypoxia. *Top,* schematic diagram of the experimental setup which allows selective induction of hypoxia at the endothelial surface. *Middle,* Diameter recordings (*D*) of a canine femoral artery segment (precontracted with norepinephrine, *NE*) with and without endothelium. Only in the presence of the endothelium is a dilation observed. *Bottom,* Changes of luminal pO_2 as recorded in the inflow cannula

(Busse et al. 1984; Pohl and Busse 1989). In rat tail arteries, which exhibit a pronounced sensitivity to this prostanoid, PGI_2 could be identified as the main mediator if this hypoxic vasodilator response. In all other arteries studied so far, EDRF mediates the hypoxic vasodilation: EDRF-blocking substances like oxyhemoglobin, whch scavenges EDRF (Martin et al. 1985) and blocks endothelium-dependent dilation in vitro but not in vivo, abolished the hypoxia-induced dilation in these segments. Moreover, in bioassay experiments, the release of an EDRF-like vasodilator compound from vascular segments as well as from cultured endothelial cells could be demonstrated during superfusion with low pO_2 (Pohland and Busse 1989).

In contrast, no release of these two autacoids can be detected under true *anoxic* conditions. This may be explained by the fact that the synthesis of EDRF (NO) from L-arginine (Palmer et al. 1988) as well as that of PGI_2 from arachidonic acid Lands et al. (1978) involves the incorporation of molecular oxygen. During true anoxia, a vasoconstriction has even been described (Borda et al. 1980; Rubanyi and Vanhoutte 1985) which is apparently not only due to the lack of formation of endothelial dilator autacoids, but also due to the release of endothelium-derived consticting factor(s) (Rubanyi and Vanhoutte 1985) as well as potassium ions and norepinephrine (Borda et al. 1980). Changes in oxygen tension not only affect the release, but also the biological half-life activity of EDRF (NO) (Busse et al. 1985). Hyperoxia results in an enhanced generation of oxygen free radicals, predominantly of superoxide anions, which inactivate EDRF (Rubanyi and Vanhoutte 1986). This effect may contribute to the reduced organ blood perfusion which is often observed under hyperoxic conditions. This vasoconstriction leads to a paradoxic impairment of tissue oxygen supply during hyperoxia (Lund et al. 1980). Recent experiments suggest, that the endothelium might generate additional vasodilator signals in response to hypoxia. In canine carotid arteries during stepwise reduction of pO_2, a progredient, endothelium-dependent hyperpolarization of vascular smooth muscle cells was demonstrated (Grote et al. 1988). Its functional significance and its dependence on the release of a recently identified endothelium-derived hyperpolarizing factor (Harder et al. 1989) is not yet clear.

Potential Mechanisms of Endothelial Oxygen Sensing

The experimental evidence is consistent with a role of endothelial cells as oxygen sensors in the vascular wall, since they respond in a pO_2-dependent way with the release of vasoactive autacoids. The primary oxygen sensing process within the endothelial cells has not yet been clarified. It has been inferred that in hypoxic cells, there is a reduction of intracellular ATP which might affect especially ATP-dependent processes at the cell membrane. In liver cells, experimental findings suggest the existence of intracellular ATP gradients and a relative lack of ATP at the cell membrane when the cellular production of ATP is reduced (Aw and Jones 1985). In endothelial cells, a reduced level of ATP could result in an increase of the PGI_2-precursor arachidonic acid, since the reacylation of this compound into membrane glycerophospholipids is dependent on ATP. In cultured bovine aortic endothelial cells, however, we observed neither a redction in oxygen consumption nor a decrease in cellular ATP content after a 30-min exposure of the cells to superfusion at a pO_2 of 30 mmHg (Fig. 2).

An alternative hypothesis is based on the observation that in the effluent of hypoxically perfused organs even increased amounts of ATP (Hopward et al. 1989) could be detected. Since several adenine nucleotides are potent stimulators of the endothelial autacoid release (Busse et al. 1988; Lückhoff et al. 1987), it is conceivable, that the enhanced release of EDRF and PGI_2 is due to the action of a hypoxia-induced release of ATP on endothelial cells. In cultured aortic endothelial cells however, the blockade of the endothelial ATP (P_2y)-receptors by reactive blue, did not affect the hypoxia-induced release of PGI_2 (Fig. 3). It remains also unclear, how hypoxia could induce pecifically the release of ATP from intact cells and what could be the main source of ATP in the hypoxic tissue.

At present, perhaps the most important finding is the observation that hypoxia induces an increase of intracellular free Ca^{2+} in cultured porcine and aortic endothelial cells (Lückhoff et al. 1986). Since the synthesis of PGI_2 and EDRF are Ca^{2+}-dependent, this increase in intracellular Ca^{2+} might repre-

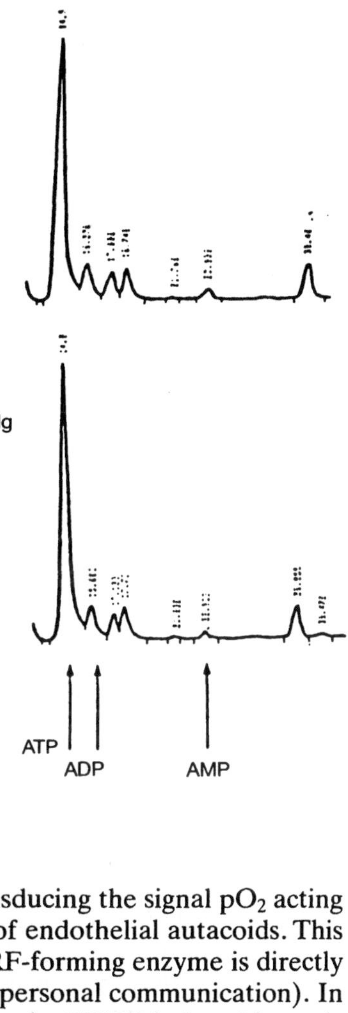

Fig. 2. Distribution of cytosolic ATP, ADP, and AMP (high-performance liquid chromatography analysis) in cultured bovine endothelial cells after 30 min superfusion with normoxic (pO_2, 140 mmHg) or hypoxic (pO_2, 25 mmHg) Tyrode's solution. 30 min hypoxia did not induce any detectable changes

sent an essential step in the signal cascade transducing the signal pO_2 acting on endothelial cells into an enhanced release of endothelial autacoids. This holds especially true for EDRF, since the EDRF-forming enzyme is directly dependent on Ca^{2+} (A. Mülsch and R. Busse, personal communication). In fact, the removal of extracellular Ca^{2+} abolishes the EDRF-induced hypoxic vasodilation of isolated arteries (Pohl and Busse 1989).

Physiologic Implications of Hypoxia-Induced Endothelial Autacoid Release

Since cultured and native endothelial cells respond to a decrease in pO_2 by an enhanced release of vasodilator autacoids, endothelial cells behave functionally as oxygen senors within the vascular wall. Due to the experimental difficulties, this endothelial oxygen sensor function has not yet been conclusively demonstrated under in vivo conditions. It is, however, interesting to note that PG appear not to be involved in the oxygen reactivity of microvessels (Jackson 1986), whereas lipoxygenase inhibitors which are also

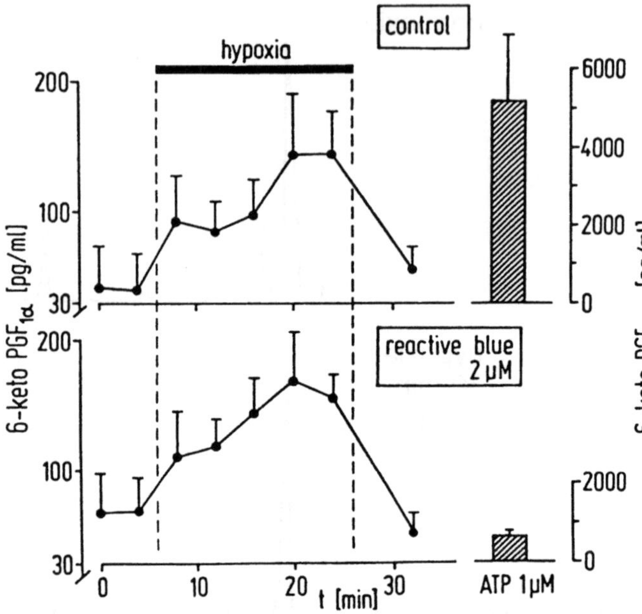

Fig. 3. Lack of effect of the P_{2y}-receptor blocking compound reactive blue on the hypoxia-induced (pO_2, 24 ± 6 mmHg) release of PGI_2 (measured as 6-keto $PGF_{1\alpha}$). Reactive blue virtually abolished the ATP-induced (*bars*), but not the hypoxia-induced release of PGI_2, indicating that hypoxic response is not due to an effect of released ATP (n = 6; x ± SEM)

unspecific inactivators of EDRP, abolished the oxygen reactivity of hamster arterioles in vivo (Jackson 1988). Since stereospecific inhibitors of the EDRF synthesis (Mülsch and Busse 1991) have now been shown to be effective in the microcirculation (Pohl et al. 1991), a more detailed investigation of th erole of EDRF in the blood flow responses to hypoxemia in vivo should be possible. The available data on in vitro experiments suggest, that endothelium-dependent dilations may significantly contribute to the local pO_2-dependent regulation of blood flow and may represent a major biologic equivalent of vascular oxygen reactivity. An endothelial oxygen sensor in the vascular wall enables the vasculature to respond rapidly to decreases in arterial pO_2 without the need for previous accumulation with the potential risk of hypoxia-induced parenchymal cell damage. Furthermore, the hypoxic signal can be effective in condult arteries whch are not reached by tissue metabolites, although they can contribute considerably to the overall vascular resistance of an organ. In addition, endothelial cells may not only respond to acute changes of oxygen tension, but may be under a constant stimulatory influence of the reduced pO_2 existing in small arteries and arterioles (Duling and Berne 1970). This may have important implications for the balance of the vasomotor tone in thee vessels and for the adjustment of tissue perfusion. Only recently, it has been shown that the inhibition of EDRF release by NG-nitro-arginine induced a significant reduction of arteriolar diameters in the hamster skin (Pohl 1991). Consistent with this, similar inhibitors of EDRF synthesis induced a hypertensive response in rabbits (Rees et al. 1989). Moreover, in rabbit hindlimb skeletal muscle, the pO_2 distribution is significantly deteriorated after blockade of EDRF release (Pohl and Busse 1988) and obviously, the expected accumulation of tissue metabolites is unable to compensate for the impairment of nutritive tissue perfusion.

Conclusion

In conclusion, recent in vivo findings suggest a decisive role of endothelial cells in the modulation of vascular tone in resistance vessels. Since EDRF synthesis and stability are pO_2-dependent, the endothelium may well represent an effective vascular oxygen sensing system which responds to acute and chronic changes of the pO_2 of the blood with potent vasoactive signals.

Acknowledgement. The author thanks Dr. A. Ogilvie, Erlangen, for the HPLC-determinations of cytosolic nucleotide fractions. This work was supported by the DFG (Po 307/1–7).

References

Aw TY, Jones DP (1985) ATP concentration gradients in cytosol of liver cells during hypoxia. Am J Physiol 249:C385–C392
Bassenge E, Busse R (1988) Endothelial modulation of coronary tone. Prog Cardiovasc Dis 30:349–380
Bauer C, Kurtz A (1989) Oxygen sensing in the kidney and its relation to erythropoietin production. Annu Rev Physiol 51:845–856
Berne RM, Rubio R (1977) Circulatory effects of tissue oxygen tension sensors. In: Reivich M, Coburn R, Lahiri S, Chance B (eds) Tissue hypoxia and ischemia. Plenum, New York, pp 163–174
Bevilacqua MP, Gimbrone MA Jr (1987) Inducible endothelial functions in inflammation and coagulation. Semin Thromb Hemost 13:425–433
Borda LJ, Shuchleib R, Henry PD (1980) Hypoxic contraction of isolated canine coronary artery. Mediation by potassium-dependent exocytosis of norepinephrine. Circ Res 46:870–879
Busse R, Förstermann U, Matsuda H, Pohl U (1984) The role of prostaglandins in the endothelium-mediated vasodilatory response to hypoxia. Pflügers Arch 401:77–83
Busse R, Trogisch G, Bassenge E (1985) The role of endothelium in the control of vascular tone. Basic Res Cardiol 80:475–490
Busse R, Ogilvie A, Pohl U (1988) Vasomotor activity of diadenosine triphosphate and diadenosine tetraphosphate in isolated arteries. Am J Physiol 254:H828–H832
Corburn RF, Grubb B, Aronson RD (1979) Effect on cyanide on oxygen tension-dependent mechanical tension in rabbit aorta. Circ Res 44:368–378
Detar R (1980) Mechanism of physiological hypoxia-induced depression of vascular smooth muscle contraction. Am J Physiol 238:H761–H769

DiCorleto PE (1984) Cultured endothelial cells produce multiple growth factors for connective tissue cells. Exp Cell Res 153:167–172
Duling BR, Berne RM (1970) Longitudinal gradients in periarteriolar oxygen tension. A possible mechanism for the participation of oxygen in local regulation of blood flow. Circ Res 27:669–678
Eyzaguirre C, Fidone SJ (1980) Transduction mechanisms in carotid body: glomus cells, putative neurotransmitters, and nerve endings. Am J Physiol 239:C135–C152
Förstermann U, Hertting G, Neufang B (1984) The importance of endogenous prostaglandins other than prostacyclin for the modulation of contractility of some rabbit blood vessels. Br J Pharmacol 81:623–630
Furchgott RF (1983) Role of endothelium in responses of vascular smooth muscle. Circ Res 53:557–573
Grote J, Siegel G, Zimmer K, Adler A (1988) The influence of oxygen tension on membrane potential and tone of canine carotid artery smooth muscle. Adv Exp Med Biol 222:481–
Harder DR, Sanchezferrer C, Kauser K, Stekiel WJ, Rubanyi GM (1989) Pressure releases a transferable endothelial contractile factor in cat cerebral arteries. Circ Res 65:193–198
Harlan JM (1987) Consequences of leukocyte-vessel wall interactions in inflammatory and immune reactions. Semin Thromb Hemost 13:434–444
Hopwood AM, Lincoln J, Kirkpatrick KA, Burnstock G (1989) Adenosine 5′-triphosphate, adenosine and endothelium-derived relaxing factor in hypoxic vasodilatation of the heart. Eur J Pharmacol 165:323–326
Jackson WF (1986) Prostaglandins do not mediate arteriolar oxygen reactivity. Am J Physiol 250:H1102–H1108
Jackson WF (1988) Lipoxygenase inhibitors block O_2 responses in hamster cheek pouch arterioles. Am J Physiol 255:H711–H716
Jackson WF, Dulling BR (1983) The oxygen sensitivity of hamster cheek pouch arterioles. In vitro and in situ studies. Circ Res 53:515–526
Lands WEM, Sauter J, Stone GW (1978) Oxygen requirement for prostaglandin biosynthesis. Prostaglandins Med 1:117–120
Longhurst J, Zells R (1979) Cardiovascular responses to local hindlimb hypoxemia: relation to the exercise reflex. Am J Physiol 237:H359–H365
Lückhoff A, Pohl U, Busse R (1986) Increased free Calcium in endothelial cells in response to hypoxia and restitution of normoxia. Pflügers Arch, Suppl 1, 406:R46
Lückhoff A, Busse R, Winter I, Bassenge E (1987) Characterization of vascular relaxant factor released from cultured endothelial cells. Hypertension 9:295–303
Lund N, Jorfeldt L, Lewis DH (1980) Skeletal muscle oxygen pressure fields in healthy human volunteers. A study of the normal state and the effects of different arterial oxygen pressures. Acta Anaesthesiol Scand 24:272–278
Martin W, Villani GM, Jothianandan D, Furchgott RF (1985) Selective blockade of endothelium-dependent and glyceryl trinitrate-induced relaxation by hemoglobin and by methylene blue in the rabbit aorta. J Pharmacol Exp Ther 232:708–716
Moncada S, Vane JR (1979) Pharmacology and endogenous roles of prostaglandin endoperoxides, thromboxane A2 and prostacyclin. Pharmacol Rev 30:293–331
Mülsch A, Busse R (1991) NG-nitro-L-arginine (N5-[imino(nitroamino)methyl]-L-ornithine) impair endothelium-dependent dilations by inhibiting cytosolic nitric oxide synthesis from L-arginine. Naunyn-Schmiedebergs Arch Pharmacol 341:143–147
Mülsch A, Böhme E, Busse R (1987) Stimulation of soluble guanylate cyclase by endothelium-derived relaxing factor from cultured endothelial cells. Eur J Pharmacol 135:247–250
Palmer RMJ, Ferrige AG, Moncada S (1987) Nitric oxide release accounts for the biological activity of endothelium-derived relaxing factor. Nature 327:524–526
Palmer RMJ, Ashton DS, Moncada S (1988) Vascular endothelial cells synthesize nitric oxide from L-arginine. Nature 333:664–666

Pearson JD, Carleton JS, Gordon JL (1980) Metabolism of adenine nucleotides by ectoenzymes of vascular endothelial and smooth muscle cells in culture. Biochem J 190:421–429
Pohl U, Busse R (1988) Reduced nutritional blood flow in autoperfused rabbit hindlimbs following inhibition of endothelial vasomotor function. In: Halpern W, Pegram B, Brayden J, Mackey K, McLaughlin M, Osol G (eds) Resistance arteries. Perinatology Press, Ithaca, pp 10–16
Pohl U, Busse R (1989) Hypoxia stimulates the release of endothelium-derived relaxant factor (EDRF). Am J Physiol 256:H1595–H1600
Pohl U, Busse R, Kessler M (1982) Vascular resistance and tissue pO_2 in skeletal muscle during perfusion with hypoxic blood. In: Kenner T, Busse R, Hinghofer-Szalkay H (eds) Cardiovascular system dynamics. Plenum Press, New York, pp 521–530
Pohl U (1991) Physiologic significance of endothelium-derived nitric oxide in the control of vascular resistance and tissue perfusion. In: Messmer K (ed) Capillary functions and white cell interaction. Prog Appl Microcirc. Karger, Basel, vol 18, pp 35–45
Rees DD, Palmer RM, Moncada S (1989) Role of endothelium-derived nitric oxide inthe regulation of blood pressure. Proc Natl Acad Sci USA 86:3375–3378
Rubanyi GM, Vanhoutte PM (1985) Hypoxia releases a vasoconstrictor substance from the canine vascular endothelium. J Physiol (Lond) 364:45–56
Rubanyi GM, Vanhoutte PM (1986) Superoxide anions and hyperoxia inactivate endothelium-derived relaxing factor. Am J Physiol 250:H822–H827
Stern DM, Carpenter B, Nawroth PP (1986) Endothelium and the regulation of coagulation. Pathol Immun 5:29–36
Whatley RE, Zimmerman GA, McIntyre TM, Taylor R, Prescott SM (1987) Production of platelet-activating factor by endothelial cells. Semin Thromb Hemost 13:445–453
Yanagisawa M, Kurihara H, Kimura S, Tomobe Y, Kobayashi M, Mitsui Y, Yazaki Y, Goto K, Masak T (1988) A novel potent vasoconstrictor peptide produced by vascular endothelial cells. Nature 332:411–415

Oxygen Transport in the Early Embryo

R. Baumann

This chapter provides a short review of oxygen transport conditions during the early embryonic development of the chick. Since mammalian embryos cannot be investigated in situ during the period of organogenesis, the experimental model of choice is the avian embryo. While there are differences in the boundary conditions for oxygen uptake between early mammalian and avian embryos, they share functional constraints which may have contributed to the persistence of the specific embryonic hemoglobins. Red blood cells and the circulatory system are the first structures that differentiate in the vertebrate embryo. The yolk sac is the site of embryonic erythropoiesis in higher vertebrates. It produces the first population of red cells (primitive erythrocytes), which synthesize the specific embryonic hemoglobins. These hemoglobins are subsequently replaced by either adult or fetal hemoglobin.

By far the best investigated experimental model is the chick embryo. Visibly hemoglobinized primitive erythrocytes appear during the 2nd day of incubation, and a closed circulatory systems is established between 48 h and 53 h of incubation. From day 6 onwards, primitive red cells are replaced by definitive red cells with adult hemoglobin (Bruns and Ingram 1973). During the period of organogenesis the area vasculosa of the yolk sac is the gas exchange organ.

The presence of specific embryonic hemoglobins has raised the question of whether they have functional properties not found in adult or fetal hemoglobin. In all species analyzed so far the alpha-chains of embryonic hemoglobin show a striking degree of homology which is much greater than to the adult alpha-chain produced in the same species (Czelusniak et al. 1982).This argues for the presence of functional constraints that have attributed to the conservation of embryonic hemoglobin structure.

Oxygen-Binding Properties of Primitive Red Cells

Primitiv red cells from the chick embryo contain four hemoglobins: the major embryonic hemoglobins P/P' (75 %), which differ only in a post-translational modification of the alpha-chain N-terminus (Engel et al. 1983), and the minor hemoglobin fractions E and M (25 %). Oxygen-binding curves of primitive red cells from 3- to 6-day-old chick embryos are polyphasic and the cooperativity of oxygen binding and the Bohr effect are saturation dependent (R. Baumann et al. 1982; Lapennas and Reeves 1983).The slope n of the Hill plot, which is the empirical index for the cooperativity of oxygen binding changes from $n=1$ at low oxygen saturation to $n=7$–8 at high oxygen saturation, i.e., the oxygen-binding curve of embryonic blood is much steeper than that of adult blood in the upper saturation range. The theoretical limit for n in a tetramer is $n=4$. Analysis of the isolated embryonic hemoglobins P and P' has shown that they form complexes of several tetramers under physiological conditions so that at least 8–12 globin chains could interact (R. Baumann and Gros 1984). The aggregation is the essential prerequisite for the high cooperativity.

It is tempting to speculate that the sites for intertetrameric contact are located on the embryonic alpha-chains, which, as stated above, are remarkably homologous in their primary structure. Further experiments have shown that the high cooperativity of the aggregates depends on the presence of additional, not yet identified allosteric effectors produced by the primitive red cells (R. Baumann und Gros 1984).

Oxygen Pressure, pH, and Oxygen Content of Embryonic Blood

In the avian embryo, oxygen diffuses through the eggshell and shell membranes into the capillary bed of the area vasculosa. The oxygenated blood returns to the embryo via the vitelline veins and enters the heart, where it is mixed with intraembryonic venous blood, i.e., there is a massive central right to left shunt. Calculation shows that at days 4 and 6, about 20 % blood from the vitelline veins is mixed with 80 % intraembryonic venous blood (Meuer and Baumann, 1988). The major artery of the embryo, the dorsal aorta supplies the intraembryonic arteries as well as the vitelline artery. Therefore, the blood gas values obtained at the origin of the vitelline artery and those of the dorsal aorta are identical. Table 1 gives values for blood oxygen pressure (pO_2), pH, and oxygen saturation (SO_2) of major intra- and extraembryonic blood vessels. The pO_2 and pH values were measured with microelectrodes after removal of the eggshell and shell membranes (Meuer and Baumann 1988; Meuer et al. 1989). Thus a small part of the diffusion resistance for oxygen is lacking. However the principal resistance to oxygen uptake is located in the surface layers of the area vasculosa and causes a fall in the pO_2 of more than 60 mm Hg. Due to the intraembryonic shunt, there is a further reduction of the oxygen pressure so that the pO_2 of the dorsal aorta is only 45 and 49 mmHg at days 4 and 6, respectively. The combination of diffusion resistance and shunt causes reduction of about 100 mmHg from ambient pO_2 of 150 mmHg to the intraembryonic arterial oxygen pressure.

It is at this point that one can evaluate the advantages of the high cooperativity of embryonic blood in counteracting the effects of the shunt.

Table 1. pH, pO_2, and SO_2 of blood in major blood vessels of 4- and 6-day-old chick embryos (measured with eggshell removed)

Location	pH	pO_2 (mmHg)	SO_2 (%)
Vitelline vein			
Day 4	8.00	64	89
Day 6	7.89	81	>95
Vitelline artery/dorsal aorta			
Day 4	7.80	45	37
Day 6	7.66	49	25
Jugular vein			
Day 4	7.64	29	22
Day 6	7.42	32	12

While the oxygen content of the blood after mixing depends only on the relative mixing volumes and their respective oxygen saturation, the resulting pO_2 is dependent on the position and steepness of the oxygen-binding curve. Two mechanisms help to reduce the inevitable fall of the pO_2: (i) the blood pH decreases by 0.2 pH units between vitelline vein and dorsal aorta, which causes a rightward shift of the oxygen equlibrium curve, and (ii) the high cooperativity of oxygen binding in the upper saturation range allows large changes of oxygen saturation with only small changes of the oxygen pressure.

Using the data for the oxygen affinity and Bohr effect of adult chicken blood (F. H. Baumann and R. Baumann 1977), we have calculated the oxygen pressure in the dorsal aorta for the pH and oxygen saturation values given in Table 1 to allow a comparison with embryonic blood. The results show that if chick embryos had adult hemoglobin, the pO_2 in the dorsal aorta would drop to 16 mmHg at day 4 and to 24 mmHg at day 6. Thus the high cooperativity of embryonic blood increases the pO_2 in the dorsal aorta by as much as 25–30 mmHg, which substantially improves the conditions for oxygen supply to the tissue. The actual oxygen content of the embryonic blood is very low due to a combination of low oxygen saturation and low hemoglobin concentration.

In Fig. 1 the in vivo oxygen-binding curves are plotted for embryonic blood at days 4 and 6, which show that intraembryonic blood oxygen content ranges from 1.8 μl blood to less than 1 μl/100 μl blood. The data support the idea that the principal function of embryonic hemoglobin is to maintain adequate gradients for oxygen diffusion both in the intra- and extraembryonic circulation. The blood leaving the embryo via the vitelline arteries has to supply the lower strata of the area vasulosa with oxygen as well as the embryonic red cells, which have a high oxygen consumption during early development (Grima et al. 1983). The rather high pH found in the vitelline veins reflects the low blood pCO_2 of about 4 mmHg and is necessary to obtain

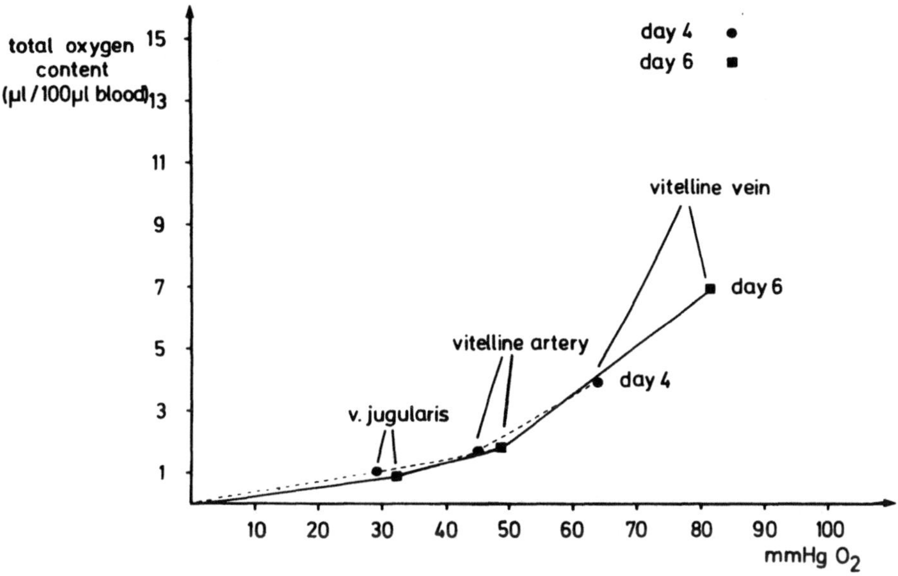

Fig. 1. In situ oxygen equilibrium curve of embryonic blood at days 4 *(circle)* and 6 *(square)* of incubation

adequate oxygenation of the embryonic blood. We found that an increase of the ambient pCO_2 to 8 mmHg decreased the pH of the vitelline vein by 0.2 pH units and reduced the oxygen saturation by about 20 % at days 4 and 6 (Meuer et al. 1989). These data in part explain the experimental result that exposure to moderate levels of carbon dioxide causes substantial embryonic mortality and malformations during early development (Romanoff 1972).

One may ask to what extent the embryo is able to adapt to hypoxia in early development. Available experimental results show that most mechanisms of defense found in the adult or during later fetal life are absent. The early embryo cannot increase its blood and red cell volume under hypoxic conditions (Table 2) nor the heart rate and minute volume (Girard 1973). Since the heart wall is not vascularized in the early embryo (cf Romanoff 1960), it is entirely dependent on the oxygen diffusing from the heart chamber into the tissue. Any decrease of the environmental pO_2 automatically reduces the oxygen supply to the heart.

These points illustrate that the oxygen transport system of the early embryo operates at the limit of its capacity. It is, therefore, no surprise that measured values for tissue oxygen pressure are close to zero (Lomholt 1984). The critical situation is also underlined by the fact that the period around days 4 and 5 of avian development shows peak values for ‚normal' ontogenetic mortality. At least 5 % of all embryos die at this stage (Romanoff 1972). In addition it has been shown that reduction of the normal ambient oxygen

Table 2. Blood volume and hemoglobin concentration of 4 to 6-day-old embryos incubated in 13.5 % oxygen and air[a]

Day	Blood volume (µl)		RBC (µl)		Hb (g/dl)	
	A	B	A	B	A	B
4	43	50	8.7	10.8	3.0	3.3
5	92	111	18.9	24.8	3.6	3.7
6	151	144	36.3	36.4	4.5	5.4

[a] RBC, red blood-cell; Hb, hemoglobin; A, 13.5 % oxygen; B, air
Blood and red cell volume were determined as described in R. Baumann et al. 1983

pressure invariably increases embryonic mortality (cf Romanoff 1972). The only adaptive change we could demonstrate in hypoxic chick embryos was a premature appearance of definitive red cells and correlated with it a higher blood oxygen affinity. These changes are, however, not found prior to day 6 of incubation (R. Baumann et al. 1983).

With regard to mammalian embryos at the same stage of development, one can assume that the loading pO_2 for oxygen uptake is lower than in the avian embryo and the pCO_2 much higher. The basic arrangement of the circulation is, however, identical to that of the avian embryo; similarly the oxygen capacity will also be low. Thus the mammalian embryo would benefit from elevated cooperativity of oxygen binding. Indeed the few analyses of mammalian embryonic blood that have examined the oxygen-binding curve over the whole range of saturation show a saturation dependence of the cooperativity and n values of 4 or higher than 4 in the upper saturation range (Brittain and Wells 1983; Holland et al. 1989). Moderate reductions of the loading pO_2 or blood pH will have the same consequences for embryonic oxygen supply as described for the avian embryo. Since the gas transport system works at the limit of its capacity at this stage, its rate of development determines embryonic survival and may serve as one major factor of embryonic selection during the period of organogenesis.

References

Baumann FH, Baumann R (1977) A comparative study of the respiratory properties of bird blood. Respir Physiol 31: 333–343

Baumann R, Gros G (1984) A comparativ study of the aggregation of avian embryonic and adult hemoglobin. In: Schnek AG, Paul C (eds) Hemoglobin. Editions de l'université de Bruxelles, pp 151–160

Baumann R, Padeken S, Haller EA (1982) Functional properties of embryonic chicken hemoglobins. J Appl Physiol 53: 1439–1448

Baumann R, Padeken S, Haller EA, Brilmayer T (1983) Effects of hypoxia on oxygen affinity, hemoglobin pattern and blood volume of early chicken embryos. Am J Physiol 244: R733–R741

Brittain T, Wells RMG (1983) Oxygen transport in early mammalian development:molecular physiology of embryonic hemoglobins. In: Johnson MH (ed) Development in mammals, vol 5. Elsevier, Amsterdam, pp 136–154
Bruns GAP, Ingram VM (1973) The erythroid cells and hemoglobins of the chick embryo. Philos Trans R Soc London [Biol] 266: 225–305
Czelusniak J, Goodmann M, Hewett-Emmett D, Weiss ML, Venta PJ, Tashian RE (1982) Phylogenetic origins and adaptive evolution of avian and mamalian hemoglobin genes. Nature 298: 297–300
Engel JD, Rusling DJ, McCune KC, Dodgson JB (1983) Unusual structure of the chicken embryonic alpha-globin gene. Proc Natl Acad Sci USA 80: 1392–1396
Girard H (1973) Arterial pressure in the chick embryo. Am J Physiol 224: 454–460
Grima M, Girard H, Dejours P (1983) Blood oxygen consumption and erythrocyte types in embryonic and postnatal chicken. Am J Physiol 244: C32–C36
Holland RAB, Rimes AF, Comis A, Tyndale-Biscoe CH (1989) Oxygen carriage and carbonic anhydrase activity in the blood of a marsupial, the tammar wallaby (Macropus eugenii), during early development. Respir Physiol 73: 69–86
Lapennas GN, Reeves RB (1983) Oxygen equilibrium curve shape in blood of chicken embryos. Respir Physiol 52: 13–26
Lomholt JP (1984) A preliminary study of local oxygen tensions inside birds eggs and gas exchange during early stages of embryonic development. In: Seymour RS (ed) Respiration and metabolism of embryonic vertebrates. Junk, Dordrecht, pp 289–297
Meuer HJ, Baumann R (1988) Oxygen pressure in intra- and extraembryonic blood vessels of early chick embryo. Respir Physiol 71: 331–332
Meuer HJ, Sieger U, Baumann R (1989) Measurement of pH in blood vessels and interstitium of 4 and 6 day old chick embryos. J Dev Physiol 11: 354–359
Romanoff AL (1960) The avian embryo. Macmillan, New York
Romanoff AL (1972) Pathogenesis of the avian embryo. Wiley, New York

Laser Doppler Flowmetry and Fetal Oxygenation

J. G. Aarnoudse and T. M. Smits

Introduction

Due to the inaccessibility of the fetus, continuous direct monitoring of the carotid arterial blood pH and gas values is not yet feasible. Information on the fetal arterial pH and blood gas values can be obtained from the capillary circulation of the presenting part, usually the fetal scalp. As long as the scalp blood flow remains undisturbed, the peripheral blood pH and gas values will form a reasonable reflection of the carotid arterial values. However, if scalp blood flow decreases, the difference between the carotid arterial and peripheral blood pH and gas values will increase, due to local skin metabolism rendering these values questionable for fetal monitoring.

The methods for continuous blood gas monitoring, e.g., the transcutaneous measurement of pO_2 ($tcpO_2$) and pCO_2 ($tcpCO_2$), are particularly dependent on a sufficiently high blood flow rate (Lübbers 1979; Steinacker and Spittelmeister 1988). The $tcpO_2$ sensor even requires heat-induced hyperemia in the fetal scalp (Huch et al. 1977). The needle electrodes used for the continuous recording of pO_2 (Aarnoudse et al. 1981; Walker et al. 1968) and pH (Wood et al. 1978) in the subcutaneous tissue also require a sufficiently high blood flow rate in order to yield an acceptable estimation of the carotid arterial values. In the widely used intermittent fetal blood sampling technique, pH and blood gas values are determined from a drop of blood, originating from the microvasculature of the scalp (Saling 1961). This sample mainly consists of blood from the arterioles, but also of blood from the capillaries and venules; this renders the method sensitive to changes in scalp blood flow. At present, the applicability of pulse oximetry for the assessment of fetal oxygen saturation is being investigated (Mendelson and Ochs 1988; Roth et al. 1989). This method requires sufficient arterial pulsations in the microvasculature of the fetal scalp, as it derives its signal mainly from the arterioles. Presumably, it suffers less interference from a decrease in the scalp blood flow than the methods which derive their signal from the capillaries.

Although no studies have been published on fetal scalp blood flow using a method which is only sensitive to the rate of flow, there is indirect evidence that scalp blood flow is reduced during labor (O'Connor and Hytten 1979), probably as a result of pressure exerted on the fetal head (Fall et al. 1980;

O'Connor et al. 1979). It has been observed that $tcpO_2$ on the fetal scalp was almost zero for several hours preceding the birth of a vigorous and healthy infant that did not show any sign of prolonged hypoxemia (O'Connor and Hytten 1979). In the same study, there was no significant relationship between the umbilical arterial pO_2 recorded in the last 5 min before birth, presumably as a result of a decreased scalp blood flow. In another study, however, it was suggested that blood flow in the fetal scalp was not seriously impaired during labor, because no significant difference was found between the umbilical arterial and scalp pH (Bönisch and Saling 1976). On the other hand, experimental work in the fetal lamb has confirmed that moderate pressure on the fetal scalp can cause a considerable decrease in the scalp blood flow and $tcpO_2$ (Aarnoudse et al. 1983; Jansen and Nathanielsz 1980).

Theoretically, it seems likely that fetal scalp blood flow is decreased by the catecholamine increase observed during labor and during hypoxemia in the human fetus (Bistoletti et al. 1983; Nylund et al. 1979), as well as in the fetal lamb (Cohn et al. 1974; Lewis et al. 1984; Robinson et al. 1981). Changes in cardiac output which occur during decelerations in the fetal heart rate (Myers et al. 1973) and changes in fetal activity (Suichies et al. 1988) may also influence the scalp blood flow during labor. The thermoregulatory control of the skin microvasculature is not yet fully developed at birth (Ryan 1976).

So far, the question as to what happens to the scalp blood flow during labor has not been answered satisfactorily, because a suitable technique for continuous intrapartum measurement was not available. To this end, we have developed a system for continuous transcutaneous measurement of the skin blood flow which can be easily applied to the fetal scalp during labor (Aarnoudse and Smits 1982).

Laser Doppler Flowmetry (Method)

The measuring system employs the laser Doppler technique described elsewhere (Bonner and Nossal 1981; Smits and Aarnoudse 1984; Stern 1985), in which the Doppler shift of monochromatic red light from a helium-neon laser, scattered back by moving red blood cells in the microvascular bed of the skin, is used to determine a flow index (Fig. 1).

Our method was validated in laboratory experiments comprising 15 experiments, which showed pronounced changes in forehead skin blood flow during hyperventilation in the adult (Smits et al. 1987). The instrument showed a reproducible linear response to the red blood cell flow in 15 small, approximately parallel, transparent tubes of 100 μm diameter embedded in opague material when the red blood cell velocity was varied from 0.2 to 5.0 mm/s. In vitro, the gain and position of the instrument showed a negligible drift. As red light from a helium-neon laser is used, changes in hemoglobin absorptivity, as occur during hypoxemia, are likely to influence the amplitude of the laser Doppler signal. Reducing the hemoglobin oxygen saturation from

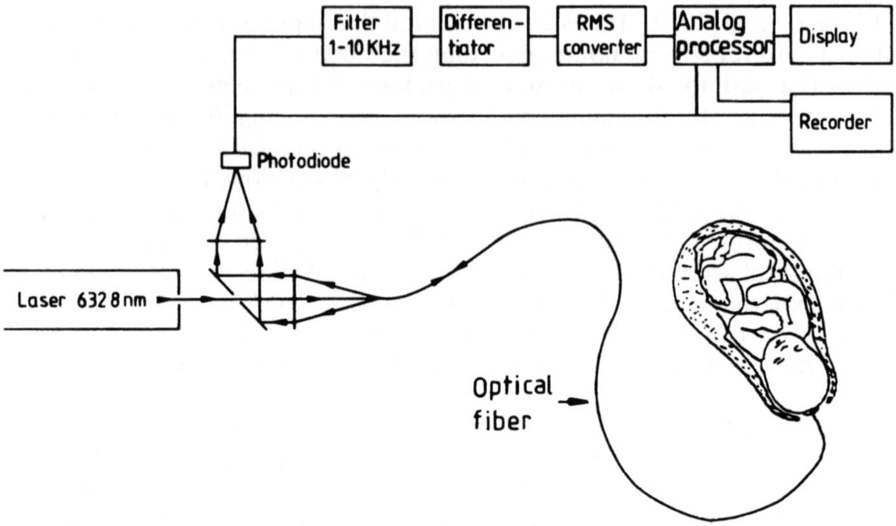

Fig. 1. Optical and electronic components of the laser Doppler system for intrapartum recording of fetal scalp blood flow. Monochromatic light from a helium-neon laser is carried to the fetal scalp by a monofilament optical fiber. In the scalp, the monochromatic light is partly scattered by moving red blood cells, which results in a Doppler shift. The backscattered light from the skin is collected by the same optical fiber and illuminates a photodetector which converts the optical signal to an electronic signal that is further processed to a flow index. The analog processor calculates the square of the RMS value of the differentiated Doppler signal

50 % to 20 % in the same microcirculatory model as was used for the linearity tests resulted in an average decrease in the laser Doppler flow (LDF) index of 4.8 %.

The laser Doppler technique has been used to record the changes in fetal scalp blood flow during the course of labor and to investigate the causes and effects of these changes in the scalp blood flow of the human fetus and the fetal lamb.

Fetal Scalp Blood Flow During Labor

With the laser Doppler technique adapted for fetal intrapartum measurement, it was possible to obtain recordings from 15 fetuses. LDF measurements were made from a cervical dilatation of 3 cm until delivery. Fetal scalp blood flow showed a decrease in the great majority of patients (Fig. 2), especially towards the end of the first stage of labour. In seven patients, scalp blood flow fell by more than 50 % during the first stage of labor. More than half of the uterine contractions recorded were accompanied by a temporary

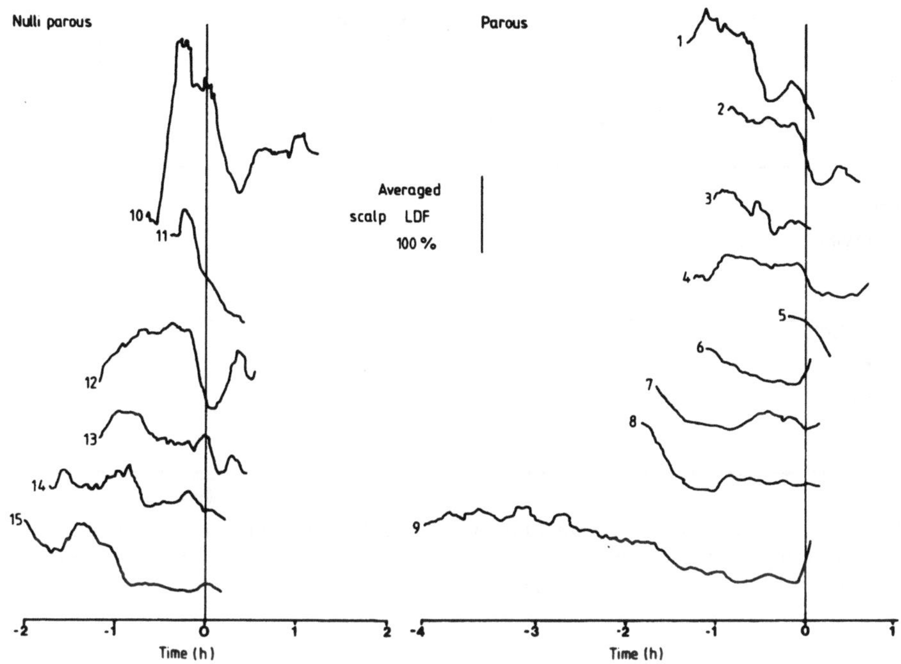

Fig. 2. The course of fetal scalp blood flow measured by the laser Doppler flow *(LDF)* technique in 15 patients during labor. The moving average of the scalp blood flow is presented as a percentage of the initial value. All recordings are synchronized at the time of full dilatation (time point 0 h). (From Smits and Aarnoudse 1984)

decrease in scalp blood flow, often by 40%–50% (Smits and Aarnoudse 1984).

LDF and $tcpO_2$ were measured simultaneously in 14 women using a combined sensor. Figure 3 shows the course of LDF and the simultaneously recorded $tcpO_2$ of one fetus during labor. Both LDF and $tcpO_2$ decrease as labor progresses. In general, the relationship between LDF and $tcpO_2$ showed a hyperbolic curve (Fig. 4): at high flow levels the $tcpO_2$ was independent of the flow rate, but at lower flow levels the pO_2 was highly dependent on the blood flow rate. The latter situation was found in most fetuses as labor progressed, indicating that under these circumstances the pO_2 no longer reflected the carotid arterial pO_2 value (Smits and Aarnoudse 1989). In two fetuses with markedly abnormal heart rate patterns, an inverse relationship between LDF and $tcpO_2$ was observed, suggesting a compensatory increase in scalp blood flow during hypoxemia (Smits et al. 1989).

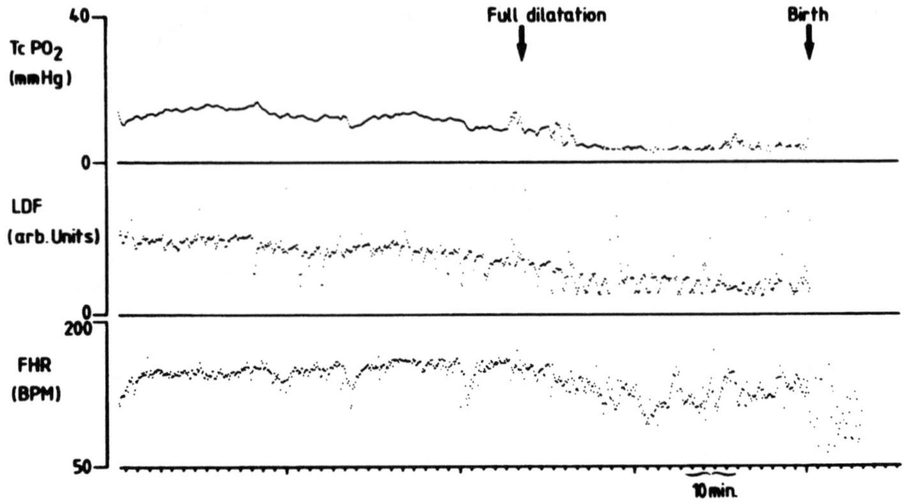

Fig. 3. The changes in laser Doppler blood flow *(LDF)* and transcutaneous *(c)* pO_2 of the fetal scalp in one patient during the entire course of labor.(*IUP*, intrauterine pressure). Changes in LDF are presented as a percentage of the initial value (From Aarnoudse and Smits 1986)

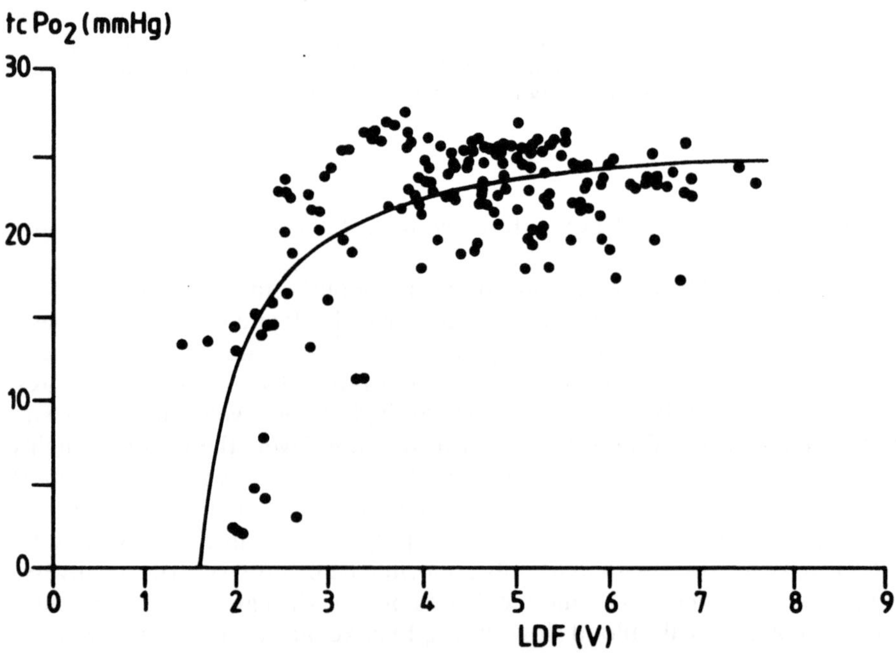

Fig. 4. Scatter diagram of 20-s mean laser Doppler flow *(LDF)* and transcutaneous *(tc)* pO_2 sample pairs of a recording during the first stage of labor. The hyperbolic curve drawn is fitted through the scatter diagram by least squares approximation. (From Smits et al. 1989)

Laser Doppler Scalp Blood Flow During Hypoxemia in the Fetal Lamb

The influence of hypoxemia on scalp blood flow was investigated in four chronically catheterized fetal lambs (125–135 days) in nine experiments. A reduction of the oxygen saturation of fetal carotid arterial blood from an average control value of 52%–21% for 20 min, resulted in an average decrease in LDF of 7% and a significant fall in the variability of scalp blood flow (Smits et al. 1986).

In another series of experiments in the ovine fetus, uterine blood flow was diminished by an inflatable balloon. This resulted in progressive severe fetal hypoxemia and acidemia, which was followed by a considerable reduction in scalp blood flow, as shown in Figs. 5 and 6. Table 1 shows the relationship between LDF and the arterial blood gas values in these experiments.

It is concluded that fetal scalp blood flow is markedly reduced during the course of labor. The increasing circular pressure on the fetal scalp seems to be one of the major causes of this decrease in scalp blood flow. Compared with the dramatic fall in scalp blood flow when circular or local pressure is exerted on the fetal scalp (Aarnoudse et al. 1983; Smits et al. 1991), moderate hypoxemia causes only a small decrease in scalp blood flow. However, severe asphyxia causes a considerable reduction in fetal scalp blood flow.

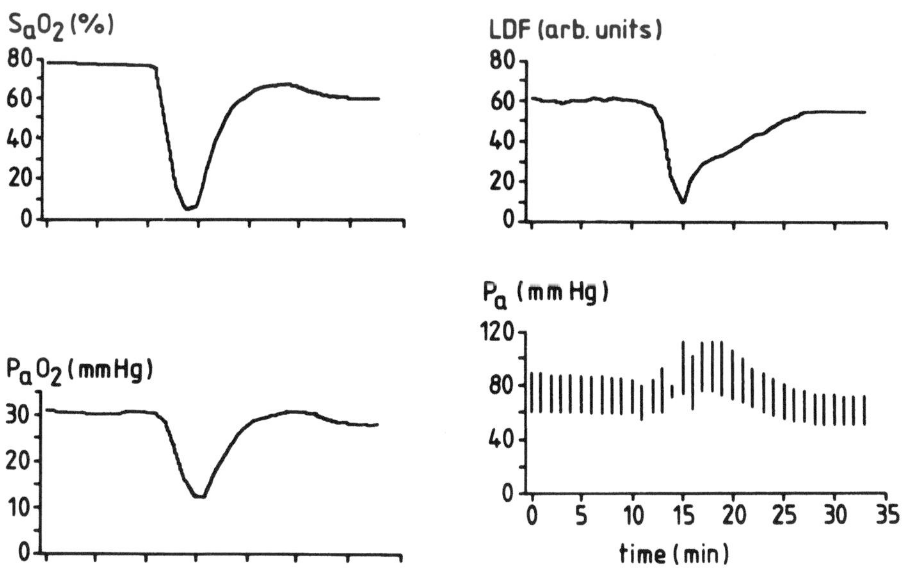

Fig. 5. Changes in laser Doppler flow *(LDF)* of the scalp, carotid arterial pO_2 and SO_2, and blood pressure in a fetal lamb during transient reduction of uterine blood flow

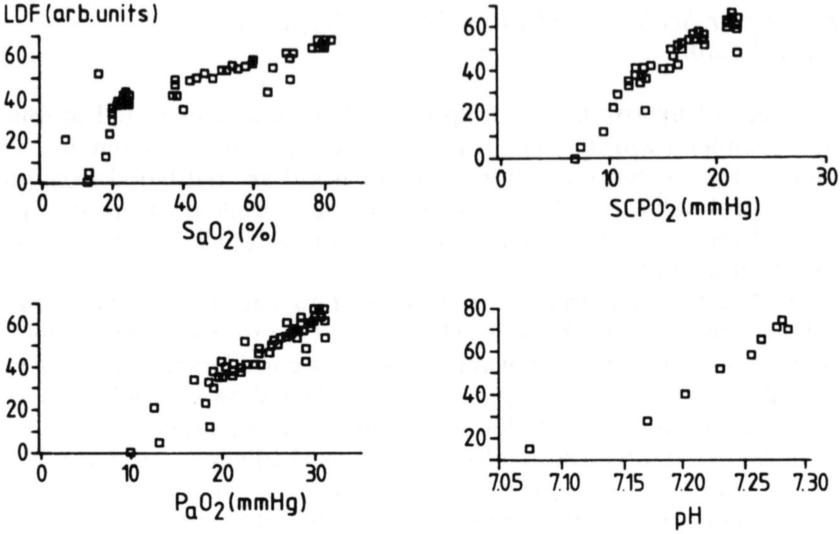

Fig. 6. The relationship between laser Doppler scalp blood flow and carotid arterial pO_2 in a fetal lamb during reduced uteroplacental blood flow

Table 1. Correlation coefficients between carotid arterial blood gas values and laser Doppler scalp blood flow during progressive hypoxemia and acidemia in fetal lambs induced by reducing uteroplacental blood flow[a]

Number of experiment	Duration (min)	SaO_2	PaO_2	pH	Number of pH samples
1	130	0.80	0.53	0.99	6
2	88	0.65	0.34	0.97	10
3	144	0.89	0.80	0.92	11
4	150	0.66	0.60	0.86	12
5	126	0.87	0.93	0.94	10

SaO_2, carotid arterial oxygen saturation; PaO_2, carotid arterial oxygen pressure
[a] Samples were taken once every 2 min; pH was determined in blood samples taken every 7.5 min.

References

Aarnoudse JG, Smits TM (1982) Continuous measurement of fetal scalp blood flow during labour by the laser Doppler technique. In: Thiery MK (ed) 8th European Perinat Congress, Brussels (abstr no 202)

Aarnoudse JG, Smits TM (1986) Laser Doppler flowmetry in the fetal scalp as related to transcutaneous blood gas measurements. In: Maas AMF et al. (eds) Physiology and methodology of blood gasses and pH. Radiometer, Copenhagen, pp 45–53

Aarnoudse JG, Huisjes HJ, Oeseburg B, Zijlstra WG (1981) Subcutaneous oxygen tension in the fetal scalp during labour. Continuous monitoring with a needle electrode. Br J Obstet Gynaecol 88: 517–524
Aarnoudse JG, Crawley JCW, Flecknell PA, Hytten FE (1983) Scalp blood flow measured by the xenon clearance technique and transcutaneous Po_2 in the fetal lamb. Pediatr Res 17: 982–985
Bistoletti P, Lagercrantz H, Lunell NO (1983) Fetal plasma catecholamine concentrations and fetal heart-rate variability during first stage of labour. Br J Obstet Gynaecol 90: 11–15
Bönisch H, Saling E (1976) The reliability of pH values in fetal blood samples. A study of the second stage. J Perinat Med 4: 45
Bonner R, Nossal R (1981) Model for laser Doppler measurement of blood flow in tissue. Appl Optics 20: 2097–2107
Cohn HE, Sachs EJ, Heymann MA, Rudolph AM (1974) Cardiovascular responses to hypoxemia and acidemia in fetal lambs. Am J Obstet Gynecol 120: 817–824
Fall O, Ek B, Nilsson BA, Rooth G (1980) The effects of mechanical pressure on local stasis on transcutaneous monitoring of fetal oxygen tension. Br J Obstet Gynaecol 87: 230–233
Huch A, Huch R, Schneider H, Rooth G (1977) Continuous transcutaneous monitoring of fetal oxygen tension during labour. Br J Obstet Gynaecol 84 [Suppl 1]: 4–39
Jansen CAM, Nathanielsz PW (1980) Pressure related changes in fetal transcutaneous pO_2 measurement. J Physiol (Land) 308: 28P
Lewis AB, Wolf WJ, Sischo W (1984) Cardiovascular and catecholamine responses to successive episodes of hypoxemia in the fetus. Biol Neonate 45: 105–111
Lübbers DW (1979) Cutaneous and transcutaneous pO_2 and pCO_2 and their measuring conditions. Birth Defects 15: 13–31
Mendelson Y, Ochs BD (1988) Noninvasive pulse oximetry utilizing skin reflectance photoplethysmography. IEEE Trans Biomed Eng 35: 798–805
Myers RE, Müller-Heubach E, Adamson K (1973) Predictability of the state of fetal oxygenation from a quantitative analysis of the components of late deceleration. Am J Obstet Gynecol 15: 1083–1094
Nylund L, Lagercrantz H, Lunell NO (1979) Catecholamines in fetal blood during birth in man. J Dev Physiol 1: 427–430
O'Connor MC, Hytten FE (1979) Measurement of fetal transcutaneous oxygen tension. Problems and potential. Br J Obstet Gynaecol 86: 948–953
O'Connor MC, Hytten FE, Zanelli GD (1979) Is the fetus ‚scalped' during labour? Lancet ii: 947–949
Robinson JS, Jones CT, Thorburn GD (1981) The effects of hypoxaemia in fetal sheep. J Clin Pathol 30 [Suppl 2]: 127–133
Roth DM, Druzin ML, Goodman DE (1989) Intrapartum fetal pulse oximetry. 36th Meeting of the Society of Gynecological Investigation (abstr no 366), Baltimore, USA, March 21, 1989
Ryan TJ (1976) The blood vessels of the skin. J Invest Dermatol 67: 110–118
Saling E (1961) Neue Untersuchungsmöglichkeiten des Kindes unter der Geburt (Einführung und Grundlagen). Geburtshilfe Frauenheilkunde 21: 905
Smits TM, Aarnoudse JG (1984) Variability of fetal scalp blood flow during labour: continuous transcutaneous measurement by the laser Doppler technique. Br J Obstet Gynaecol 91: 524–531
Smits TM, Aarnoudse JG, Zijlstra WG (1989) Fetal scalp blood flow as recorded by laser Doppler flowmetry and transcutaneous pO_2 during labour. Early Hum Dev 20: 109–124
Smits TM, Aarnoudse JG, Zijlstra WG (1986) Red blood cell flow in the fetal scalp during hypoxemia in the chronic sheep experiment: a laser Doppler flow study. Pediatr Res 20: 407–410

Smits TM, Aarnoudse JG, Geerdink JJ, Zijlstra WG (1987) Hyperventilations induced changes in periodic oscillations in forehead skin blood flow measured by laser Doppler flowmetry. Int J Microcirc 6: 149–159
Smits TM, Aarnoudse JG, Zijlstra WG (1990) Scalp blood flow, measured by laser Doppler flowmetry and transcutaneous pO_2 and pcO_2 in the lamb. Pediatr Res 27: 442–444
Steinacker JN, Spittelmeister W (1988) Dependence of transcutaneous O_2 partial pressure on cutaneous blood flow. J Appl Physiol 64: 21–25
Stern MD (1985) Laser Doppler velocimetry in blood and multiple scattering fluids: theory. Appl Optics 24: 1968–1986
Suichies HE, Aarnoudse JG, Okken A, Jentink HW, Mul FFM de, Greve J (1988) Forehead skin blood flow in normal neonates during active and quiet sleep, measured with a diode laser Doppler instrument. Acta Pediatr Scand 77: 220–225
Walker A, Philips L, Powe L, Wood C (1968) A new instrument for the measurement of tissue pO_2 of human fetal scalp. Am J Obstet Gynecol 100: 63–71
Wood C, Anderson I, Reddy S, Shekleton P (1978) Continuous measurement of tissue pH in the human fetal scalp. Br J Obstet Gynaecol 85: 668–667

Abstracts

*Blood Flow to the Yolk Sac Placenta as a Function of Arterial Oxygen Content**

A. M. CARTER, and A. DETMER

Introduction

The yolk sac shrivels and disappears early in human development, but in other mammals it is retained until term and participates in fetomaternal exchange. The vitelline vessels which supply the inverted yolk sac placenta of the guinea pig pursue a different course from the blood vessels that supply the chorioallantoic placenta. The vitelline artery can, therefore, be catheterized without interrupting the umbilical blood flow, and can be used to obtain a reference sample of blood when determining blood flow to the lower body and placenta by the microsphere technique (Carter 1984). Evidently, tying a catheter into the vitelline artery precludes determination of blood flow to the yolk sac placenta. An alternative approach, which we have used when measuring cerebral blood flow (Carter and Gu 1988), is to catheterize the axillary artery. However, because of preferential streaming of blood and microspheres in the fetal circulation, a reference sample from a branch of the ascending aorta cannot be used to measure blood flow in organs supplied from the descending aorta. Blood flow to the yolk sac placenta of the guinea pig has, therefore, been estimated by an indirect approach.

Methods

Microspheres were injected in the saphenous vein of seven fetuses, and the ratio between the radioactivity in the yolk sac ($CPM_{yolk\ sac}$) and the radioactivity in the gut determined. This ratio, which is the same as that between yolk sac blood flow ($\dot{Q}_{yolk\ sac}$) and gut blood flow, was then multiplied by a reference value for blood flow to the gut, as follows:

$$\dot{Q}_{yolk\ sac} = CPM_{yolk\ sac}/CPM_{gut} \times \dot{Q}_{gut}$$

The reference value for gut blood flow (2.50 ml/min) was taken from a separate microsphere study in which the vitelline artery had been catheterized to obtain a reference blood sample (Carter 1984). Arterial oxygen content and pH were measured in blood taken from an axillary artery.

* This study was supported by the Danish Medical Research Council

Results and Discussion

A positive linear correlation was found between estimated yolk sac blood flow and arterial O_2 content ($r = 0.90$; $p<0.01$) (Fig. 1). Blood flow to the gastrointestinal tract, which was used as a reference for these estimates, shows little variation with arterial O_2 content in the 2–6 m*M* range (Peeters et al. 1979). It may, therefore, be concluded that yolk sac blood flow decreases when the arterial O_2 content falls.

In normoxemic guinea pig fetuses near term, vitelline blood flow was 100–300 µl/min, which is equivalent to 0.3% – 1.0% of the biventricular cardiac output. This must be regarded as a minimum estimate, since two incisions were made in the yolk sac to obtain access to vessels in a fetal forelimb and a fetal hindlimb. Although these incisions were far from the main vitelline vessels, there was some loss of fluid, and this may have affected yolk sac function.

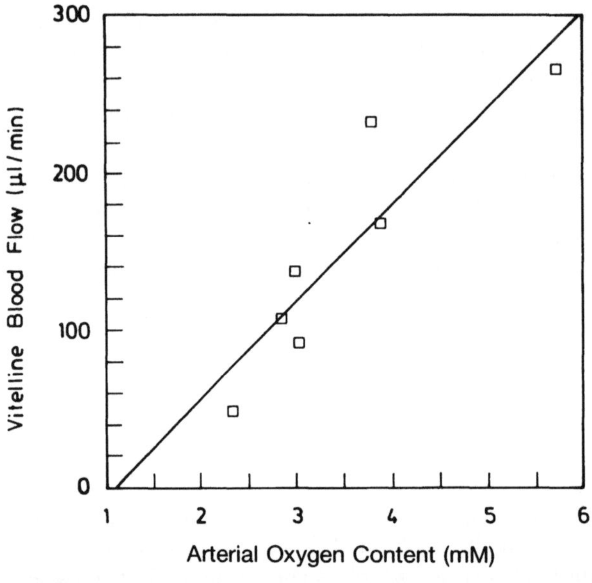

Fig. 1. Blood flow to the yolk sac placenta of the guinea pig as a function of fetal arterial oxygen content ($r = 0.90$; $p<0.01$)

References

Carter AM (1984) The blood supply to the abdominal organs of the fetal guinea-pig. J Dev Physiol 6: 407–416

Carter AM, Gu W (1988) Cerebral blood flow in the fetal guinea-pig. J Dev Physiol 10: 123–129

Peeters LLH, Sheldon RE, Jones MD, Makowski EL, Meschia G (1979) Blood flow to fetal organs as a function of arterial oxygen content. Am J Obstet Gynecol 135: 637–646

IV Chronic Oxygen Deficiency During Pregnancy: Impact on the Fetus

Development of Chemoreflexes Affecting Breathing: Influence of Chronic Hypoxia

M. A. Hanson, B. A. Williams, and P. Kumar

Introduction

The effects of acute hypoxia on fetal breathing (see Dawes 1984) and neonatal breathing (see Haddad and Mellins 1984; Rigatto 1984) are well established. This information comes from the effects of exposure of the fetus or neonate to an episode of hypoxia lasting from a few minutes to about 1 h. It is of relevance to the understanding of the processes operating during a brief hypoxic or asphyxic epidode, e.g. during the reductions in uteroplacental blood flow occurring with the uterine contractions of labour or during a period of apnoea in the neonate. This knowledge may, however, not shed much light on the processes operating during chronic or repeated acute hypoxia, e.g. in certain types of fetal growth retardation, in right to left shunting (e.g. when the ductus arteriosus remains patent) or in recurrent apnoea post-natally. The processes operating during acute and chronic hypoxia may be utterly different. The question of fetal adaptation during chronic hypoxia and its clinical relevance have been discussed recently (see Dawes et al. 1990), so in this review we will concentrate on the effects of chronic hypoxia on chemoreflexes acting on breathing. This is not to underestimate the importance of chemoreflexes acting on the circulation under such circumstances, but our emphasis arises from the relative lack of knowledge of that area (but see Bocking et al. 1986, 1988 a, b).

Chemoreflex Effects on Fetal Breathing?

The idea of chemoreflex effects on fetal breathing movements (FBM) is not new, although it assumes that the peripheral arterial chemoreceptors are functional in utero. It was, however, some time before it was shown that the carotid and aortic chemoreceptors are indeed active in fetal sheep in the second half of gestation, and that they respond to a reduction in arterial pO_2 (Blanco et al. 1984 a). Despite this finding, it is not possible to stimulate FBM in late gestation by stimulating the peripheral chemoreceptors with hypoxia (Boddy et al. 1974) or pharmacological agents (see Hanson 1990). Hypercapnia does stimulate FBM, but its effect can be over-ridden by the inhibitory effect of hypoxia, which is not consistent with an effect of CO_2 predominantly via the peripheral chemoreceptors. Finally, after chemodenervation, the incidence of FBM is reported to be unchanged (Moore et al., 1989; Rigatto et al. 1988) or only slightly reduced (Murai et al. 1985), but hypoxia still inhibits FBM. The only circumstances under which hypoxia produces an effect reminiscent of a chemoreflex by post-natal criteria is following transection of the brain stem at the upper pons, when hypoxia stimulates FBM (Gluckman and Johnston 1987). This stimulation of FBM has been observed similarly in chemodenervated transected fetuses (Dawes et al. 1983; Moore et al. 1989), although this has recently been challenged for fetuses with pontine lesions (Johnston 1991).

Chemoreflexes acting on FBM have thus yet to be established, and much attention is currently directed at the mechanisms in the fetal brain which mediate the effects of hypoxia on FBM (see Noble 1991).

Effects of Chronic Hypoxia on Fetal Breathing

Whilst FBM cease initially after reduction of arterial pO_2 from approximately 25 mm Hg to 12–15 mm Hg, the incidence of FBM returns to control over 3–24 h if the hypoxia is maintained: this is true whether the hypoxia is produced by anaemia (Bissonnette and Hohimer, 1987), maternal common iliac artery occlusion (Bocking et al. 1988 a, b) or reduced maternal inspired oxygen fraction (Koos et al. 1988). The mechanisms underlying this return are not known, but decreases in peripheral chemoreceptor sensitivity have to be considered. Exposure of the fetal chemoreceptors to hyperoxia in utero appears to induce resetting of the position of their stimulus-response curve in a manner similar to that which occurs post-natally (Blanco et al. 1988). As shown in Fig. 1, this shift results in an *increase* in their sensitivity to changes in pO_2 from any mean pO_2. Conversely, chronic hypoxia in the neonate appears to *reduce* sensitivity (see below).

Chemoreflex Effects on Neonatal Breathing

Much research has been conducted into the role played by peripheral chemoreceptors in the ventilatory response of the neonate to hypoxia (see Walker 1984; Hanson 1986 for reviews). In the newborn of every species examined, the rise in ventilation which occurs over the first few minutes of exposure to hypoxia is not sustained, and ventilation then falls towards its pre-hypoxic level. The size of the initial stimulation of ventilation increases in the early neonatal period, and direct recordings from chemoreceptor afferent fibres indicate that this is likely to be due to the resetting which occurs over this period. Such direct recordings have also revealed that the chemoreceptors do not fail or adapt as ventilation falls during acute hypoxia (Schwieler 1968; Blanco et al. 1984 b), and it appears that central nervous system (CNS) processes are involved in producing this secondary fall (Lawson and Long

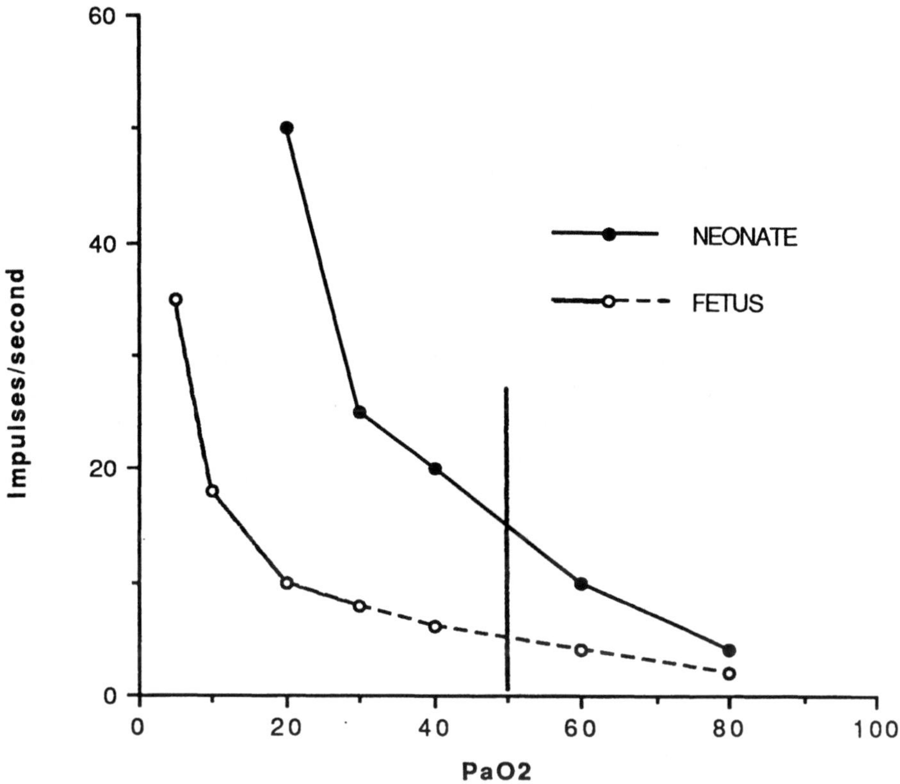

Fig. 1. Arterial chemoreceptor response curves showing position in fetal sheep near term (*broken line*, PaO_2 range after delivery) and lamb on postnatal day 10. At any PaO_2 (e.g. 50 mm Hg as shown), slope is greater after resetting, thus response to changes in PaO_2 is greater

1983; Martin-Body and Johnston 1988). It is not known whether they are the same mechanisms as those involved in producing the cessation of FBM in hypoxia. It should however be stressed that, as the hypoxia continues, these processes may alter the *gain* of the chemoreflex loop. As hypoxia continues, this change in gain is shown for example by a reduction in the transient fall in ventilation which occurs on breathing two breaths of 100% O_2 (Carroll and Bureau 1987). This is not evidence of failure or adaptation of the chemoreceptor afferent input during hypoxia.

Effects of Chronic Hypoxia in the Neonate

Several lines of evidence indicate that the ventilatory response of the neonate to an episode of acute hypoxia is attenuated by exposure to chronic hypoxia.

They include studies of infants born at high altitude (Lahiri et al. 1978), of children chronically hypoxaemic due to tetralogy of Fallot (Sorensen and Severinghaus 1968) and of rat pups (Eden and Hanson 1987) or kittens (Hanson et al. 1989 a, b) which were born and reared in a normobaric hypoxic chamber. From the discussion above, it should be clear that alteration in peripheral chemoreceptor sensitivity and/or in CNS mechanisms mediating the ventilatory response to hypoxia, including those controlling the gain of respiratory chemoreflexes, might be invoked to explain this effect of chronic hypoxia.

Figure 2 shows that chronic hypoxia alters the response in instantaneous ventilation of kittens when they are given alternate breaths of air and a hypoxic gas. Such rapid changes in inspired, and hence in arterial, pO_2 will be detected by the arterial chemoreceptors and changes in their discharge are known to change respiratory timing and volumes on a breath-to-breath basis (Cunningham et al. 1986). Chronic hypoxia also attenuates the transient fall in ventilation produced by giving two breaths of 100 % O_2 and the respiratory compensation which follows a transient disturbance in arterial blood gases produced by a spontaneous sigh (Williams et al. 1988).

These effects of chronic hypoxia to reduce chemoreflexes could theoretically be via changes in the arterial chemoreceptor afferent limb, the CNS, or in the efferent limb of the reflex. Part of the explanation may well lie in the chemoreceptors, for we found that carotid chemoreceptor hypoxic sensitivity was much reduced in chronically hypoxic kittens at 12–23 days of age (Hanson et al. 1989 a, b). We did not find this is rats, although in them it was only possible to make whole-nerve recordings at 5–10 weeks post-natally (Eden and Hanson 1987).

To address the question of the part played by CNS mechanisms, we compared the responses to acute hypoxic challenges of normoxic and chronically hypoxic rat pups at 4–7 days of age, when intact, after decerebration, or after transection of the brain stem through the upper pons (Williams and Hanson 1989). After decerebration, an initial stimulation and a secondary fall in ventilation still occurred, as in intact rats. Figure 3 shows that transection abolished the secondary fall in ventilation in both groups. These results confirm those of Martin-Body and Johnston (1988) for newborn rabbits and Martin-Body (1988) for adult rats, and emphasise the role of CNS mechanisms in producing the fall in ventilation in hypoxia. However in our experiments, the response of the chronically hypoxic animals was substantially less than that of the normoxic group even after transection of the brain stem. It, therefore, appears that the mechanism underlying the attenuation of the chemoreflux is likely to lie in the peripheral chemoreceptors.

There is much more which we need to know before we can understand the effects of chronic hypoxia on fetal and neonatal breathing. We need to establish the processes by which chronic hypoxia alters peripheral chemoreceptor function. Secondly, we need to determine the mechanisms in the fetal brain which mediate the response to hypoxia, including the part played by

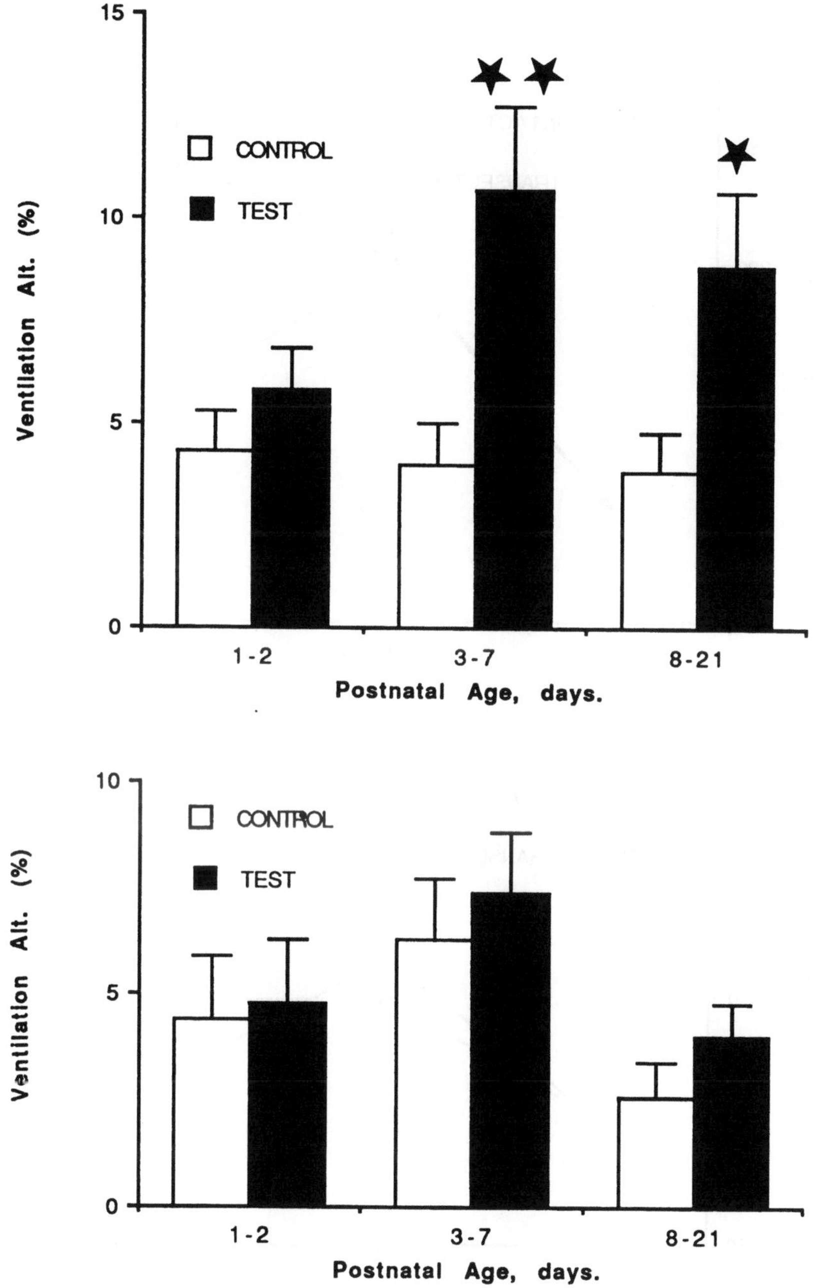

Fig. 2. The breath-to-breath alternation in ventilation produced by alternate breaths of air and 14 % O_2, compard with control runs, when air was supplied to each breath, in kittens at three postnatal ages. Significant alternations (**, $p<0.02$; *, $p<0.05$) in ventilation are present in normoxic kittens *(upper graph)* by days 3–7. Chronically hypoxic kittens *(lower graph)* do not show significant alternations at any age

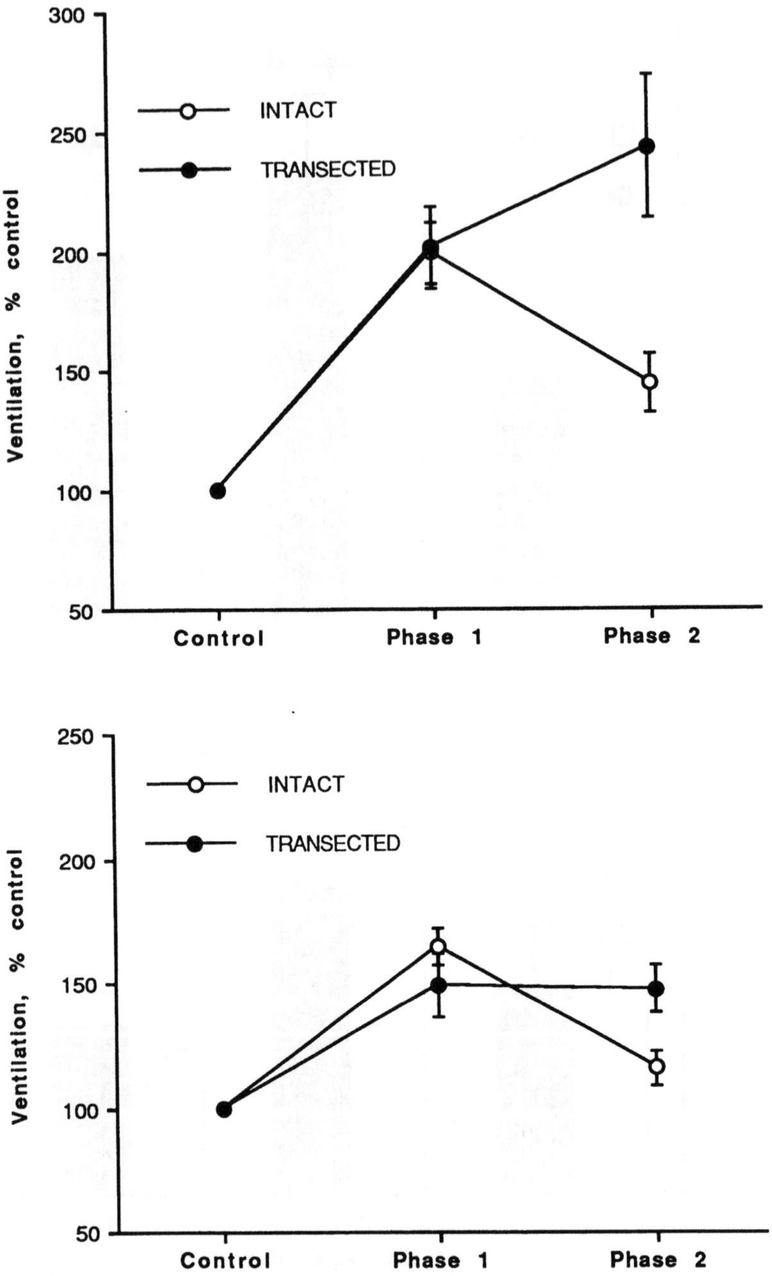

Fig. 3. The ventilatory response to acute exposure to 8 % O_2 for 6 min in rat pups aged 4–7 days. Ventilation is expressed as peak, occurring after 1–2 min *(Phase 1)* and the fall, occurring over the next 3–4 min *(Phase 2)*. The fall in phase 2 is abolished in both normoxic *(upper graph)* and hypoxic *(lower graph)* pups by transection of the brain stem through the upper pons

changes in cerebral blood flow and by changes in the release of neuromodulators.

Acknowledgements We are grateful to the Wellcome Trust, Action Research for the Crippled Child, and the Foundation for the Study of Infant Deaths for support.

References

Bissonnette JM, Hohimer AR (1987) Acute anemic hypoxia produces a transient depression in fetal respiratory activity. J Appl Physiol 63: 1942–1946

Blanco CE, Dawes GS, Hanson MA, McCooke HB (1984 a) The response to hypoxia of arterial chemoreceptors in fetal sheep and new-born lambs. J Physiol (Lond) 351: 25–37

Blanco CE, Hanson MA, Johnson P, Rigatto H (1984 b) Breathing pattern of kittens during hypoxia. J Appl Physiol 56 (1): 12–17

Blanco CE, Hanson MA, McCooke HB (1988) Effects on carotid chemoreceptor resetting of pulmonary ventilation in the fetal lamb in utero. J Dev Physiol 10: 167–174

Bocking AD, Harding R, Wickham PJ (1986) Effects of reduced uterine blood flow on accelerations and decelerations in heart rate in fetal sheep. Am J Obstet Gynecol 154: 329–335

Bocking AD, Gagnon R, Milne KM, White SE (1988 a) Behavioral activity during prolonged hypoxemia in fetal sheep. J Appl Physiol 65: 2420–2426

Bocking AD, Gagnon R, White SE, Homan J, Milne KM, Richardson BS (1988 b) Circulatory responses to prolonged hypoxemia in fetal sheep. Am J Obstet Gynecol 159: 1418–1424

Boddy K, Dawes GS, Fisher R, Pinter S, Robinson JS (1974) Foetal respiratory movements, electrocortical and cardiovascular responses to hypoxemia and hypercapnia in sheep. J Physiol (Lond) 243: 599–618

Carroll JL, Bureau MA (1987) Decline in peripheral chemoreceptor excitatory stimulation during acute hypoxia in the lamb. J App Physiol 63 (2): 795–802

Cunningham DJC, Robbins PA, Wolff CB (1986) Integration of respiratory responses to changes in alveolar partial pressures of CO_2 and O_2 and in arterial pH. In: American Physiological Society (ed) Handbook of physiology. Williams and Wilkins, Baltimore

Dawes GS (1984) The central control of fetal breathing and skeletal muscle movements. J Physiol (Lond) 346: 1–18

Dawes GS, Gardner WN, Johnston BM, Walder DW (1983) Breathing in fetal lambs: the effects of brain stem section. J Physiol (Lond) 335- 535–553

Dawes GS, Borruto F, Zacutti A, Zacutti A Jr (1990) Fetal autonomy and adaptation. Wiley, Chichester pp 1–186

Eden GJ, Hanson MA (1987) Effects of chronic hypoxia from birth on the ventilatory response to acute hypoxia in the newborn rat. J Physiol (Lond) 392: 11–19

Gluckman PD, Johnston BM (1987) Lesions in the upper lateral pons abolish the hypoxic depression of breathing in unanaesthetized fetal lambs in utero. J Physiol (Lond) 382: 373–383

Haddad GG, Mellins RB (1984) Hypoxia and respiratory control in early life. Annu Rev Physiol 46: 629–643

Hanson MA (1986) Peripheral chemoreceptor function before and after birth. In: Johnston BM, Gluckman PD (eds) Respiratory control and lung development in the fetus and newborn. Perinatology Press, New York, pp 311–330

Hanson MA (1990) Peripheral control of fetal breathing. In: Johnston BM, Gluckman PD (eds) Fetal physiology and medicine. Perinatology Press, Ithaca, pp 191–204

Hanson MA, Eden GJ, Nijhuis JG, Moore PJ (1989 a) Peripheral chemoreceptors and other O_2 sensors in the fetus and newborn. In: Lahiri S, Forster RE, Davies RO, Pack AI (eds) Chemoreceptors and reflexes in breathing. Oxford University Press, New York, pp 113–120

Hanson MA, Kumar P, Williams BA (1989 b) The effect of chronic hypoxia upon the development of respiratory chemoreflexes in the newborn kitten. J Physiol (Lond) 411: 563–574

Johnston BM (1991) Brainstem inhibitory mechanisms in the control of fetal breathing. In: Hanson MA (ed) The fetal and neonatal brainstem, Cambridge University Press, Cambridge, pp 21–47

Koos, BJ, Kitanaka T, Matsuda K, Gilbert RD, Longo LD (1988) Fetal breathing adaptation to prolonged hypoxaemia in sheep. J Dev Physiol 10 (2): 161–166

Lahiri S, Brody JS, Motoyama EK, Velasquez TM (1978) Regulation of breathing in newborns at high altitude. J Appl Physiol 44: 673–678

Lawson EE, Long WA (1983) Central origin of biphasic breathing pattern during hypoxia in newborns. J Appl Physiol 55: 483–488

Martin-Body RL (1988) Brain transections demonstrate the central origin of hypoxic ventilatory depression in carotid body-denervated rats. J Physiol (Lond) 407: 41–52

Martin-Body RL, Johnston BM (1988) Central origin of the hypoxic depression of breathing in the newborn. Respir Physiol 71: 25–32

Moore PJ, Parkes MJ, Nijhuis JG, Hanson MA (1989) The incidence of breathing movements of fetal sheep in normoxia and hypoxia after peripheral chemodenervation and brain-stem transection. J Dev Physiol 11: 147–151

Murai DT, Lee CH, Wallen LD, Kitterman JA (1985) Denervation of peripheral chemoreceptors decreases breathing movements in fetal sheep. J Appl Physiol 59: 575–579

Noble R (1991) Brainstem mechanisms mediating the neonatal ventilatory response to hypoxia. In: Hanson MA (ed) The fetal and neonatal brainstem. Ed. Cambridge Univesity Press, Cambridge, pp 48–58

Rigatto H (1984) Control of ventilation the the newborn. Annu Rev Physiol 46: 661–674

Rigatto H, Hasan SU, Jansen AH, Gibson DA, Nowaczyk BJ, Cates D (1988) The effect of total peripheral chemodenervation on fetal breathing and on the establishment of breathing at birth in sheep. In: Jones CT (ed) Perinatology Press, New York, pp 613–621

Schwieler GH (1968) Respiratory regulation during postnatal development in cats and rabbits and some if its morphological substrate. Acta Physiol Scand 72 [Suppl 304]: 1–123

Sorensen SC, Severinghaus JW (1968) Respiratory insensitivity to acute hypoxia persisting after correction of tetralogy of Fallot. J Appl Physiol 25: 221–223

Walker DW (1984) Peripheral and central chemoreceptors in the fetus and newborn. Annu Rev Physiol 46: 687–703

Williams BA, Hanson MA (1989) The effect of decerebration and brain-stem transection on the ventilatory response to acute hypoxia in normoxic and chronically hypoxic newborn rats. J Physiol (Lond) 414: 25P

Williams BA, Kumar P, Hanson MA (1988) The effect of chronic hypoxia upon the ventilatory responses to spontaneous sighs in the newborn kitten. In: Jones CT (ed) Fetal and neonatal development. Perinatology Press, Ithaca, pp 319–322

*Restricted Fetal Oxygen Supply: A Cause of Intrauterine Growth Retardation?**

J. F. CLAPP III

Introduction

This chapter focuses on the interaction between oxygen and/or substrate delivery and fetal growth. It begins with a brief review of experimental data dealing with chronic reductions in oxygen and substrate delivery. This information is then used to examine the premise that fetal growth is regulated by delivery of substrate and/or oxygen. The chapter concludes with discussion of a potential mechanism which may regulate the interaction.

Animal Experimental Data

Animal Models

Three types of experimental models are used to chronically restrict oxygen and substrate delivery to the fetus. The first simply uses systemic adrenergic responses to stimuli such as maternal heat stress, hypoxia, or nutritional deprivation to induce a chronic redistribution of maternal cardiac output away from the uteroplacental bed; the second uses mechanical constriction or occlusion of the uterine vasculature to reduce delivery; and the third reduces the size of the implantation site by either ablation of the implantation site prior to conception or ligation of a fraction of the fetal placental circulation. Each mimics a specific clinical situation.

Morphometric and Functional Correlates

Chronic maternal heat stress, severe hypoxia, and nutritional deprivation all reduce oxygen and substrate delivery and have similar negative impacts on placental and fetal weight. Alexander et al. (1987) and Bell et al. (1987) have shown that the decrease in placental and fetal weight following chronic heat stress is directly proportional to the reduction in uterine and umbilical blood flow, uteroplacental oxygen consumption, and selective aspects of placental

* This work was supported in part by the Harry W. Chandler Memorial Research Fund

transfer. While a chronic reduction of maternal pO_2 to 45–50 torr produces a 15 %–20 % balanced reduction in fetoplacental weight in a variety of species (Jacobs et al. 1988a, b), a chronic reduction of maternal pO_2 to 55–60 torr did not produce a long-term reduction in uterine flow, uteroplacental oxygen uptake, or fetal weight (Kitanaka et al. 1989a, b). Unfortunately, similar maternal hemodynamic and metabolic data has not been obtained during exposure to the degree of maternal chronic hypoxia which produced significant fetoplacental growth restriction. The same is true for nutritional stress. However, in all these models, when fetal growth restriction is present, the decrease in placental weight explains more than 85 % of the variance in fetal weight.

Uterine vascular ligation, constriction, and embolization produce the classical picture of asymmetric fetoplacental growth restriction (intrauterine growth retardation, IUGR), and there is a direct linear correlation between the decrease in fetal and placental weight and the reduction in uteroplacental blood flow (Clapp et al. 1982a, Clark and Mack 1987). In the embolization model, fetal weight and morphometry at term are directly related ($r = 0.82$) to the residual volume of functional placental tissue over ranges of 300 cm^3 and 3 kg (Clapp 1989).

Both ablation of the implantation site and placental ligation also produce the classical morphometric profile of asymmetric IUGR and near-term flow measurements in the caruncIectomy model demonstrate a direct linear relationship between both uterine and umbilical blood flows and placental and fetal weight (Owens et al. 1986).

Longitudinal measurements of fetal substrate levels and uptakes are available only in the hypoxia, embolization, and constriction models. Maternal hypoxia produces a transient fetal metabolic acidosis with hyperlactacidemia that clears in 24–48 h. Thereafter, pO_2 remains depressed, lactate levels fall towards normal, glucose concentrations are normal, and the alanine concentration rises as does red cell volume and oxygen content (Jacobs et al. 1988b; Kitanaka et al. 1989b). Longitudinal fetal uptake data are not available. In both the constriction and embolization models pH, pO_2, O_2, glucose, and lactate contents remain normal until long after there is evidence of a decrease in fetal growth, flow, and substrate uptake (Clapp et al. 1981; Charlton and Johengen 1987; K. E. Clark, 1989, personal communication).

In all models studied late in the course of IUGR, the metabolic profile reflects a supply-induced growth restriction similar to that seen after birth with nutritional deprivation. This includes a decrease in oxygen, glucose, and insulin concentrations with an increase in hemoglobin, lactate, and alanine and a hormonal milieu characteristic of that seen with exogenous growth suppression (Robinson et al. 1985). At this stage, placental transfer and the fetal uptakes of oxygen and a variety of substrates are markedly reduced. Nonetheless, their proportionality to fetal weight remains unchanged, suggesting that growth at this stage is regulated by the availability of oxygen and/or substrate. Finally, in response to repetitive embolization, there are acute and chronic changes in the distribution of uterine and umbilical blood flow which result in an improvement in placental perfusion balance (Clapp 1989) which should maximize transfer capacity.

Substrate Delivery Limitations on Feto-placental Growth

General

The fact that maternal chronic hypoxia, nutritional deprivation, heat stress, uterine vascular compromise, and placental restriction or damage all result in a similar morphometric outcome suggests that the morphometric response may be initiated by a common stimulus. Unfortunately, the functional correlative data and specific perturbations necessary to demonstrate cause and effect are not available in all models. This leaves us with two potential growth-regulating stimuli that could initiate the morphometric response shared by all the models. These are: a change in the absolute delivery of one or more substances or substrates including oxygen to the placental site or, alternately, simply a change in the rate of placental perfusion.

Oxygen as the Regulator

The fact that a chronic reduction in maternal pO_2 to 55–60 torr for 3 weeks in late gestation reduced fetal pO_2 without decreasing uteroplacental oxygen uptake or producing definitive morphometric evidence of fetal growth restriction argues strongly against the idea that small changes in placental oxygen delivery modulate the rate of fetoplacental growth. On the other hand, the authors of this critical study (Kitanaka et al. 1989b) acknowledge that a large body of evidence in other animal models (Gilbert et al. 1979) and humans (Lichty et al. 1957) demonstrates that a similar reduction in maternal inspired oxygen concentration results in morphometric evidence of fetal growth restriction. Thus, while it is unlikely that oxygen delivery is a primary modulator, it does have a defined impact on fetoplacental growth that appears to vary between species and with the duration and intensity of the stimulus.

Other Substrates

Even short-term nutritional deprivation has been shown to have an impact on fetal growth rate (Mellor 1983) and is associated with a reduction in uterine blood flow and maternal concentrations of glucose and amino acids (Morriss et al. 1980). This, coupled with the fact that fetal nutritional supplementation can prevent the fetoplacental growth restriction associated with repetitive embolization of the uterine vasculature (Charlton and Johengen 1987), indicates that absolute rate of placental delivery of one or more substrates may be the primary, day-to-day, modulator of fetoplacental growth. Clearly, the principle of supply-limited growth regulation has fetal survival value, is supported by the data obtained in other models, and has its precedent in postnatal growth. However, to date, a change in substrate delivery has not been cleanly separated from a concomitant reduction in uteroplacental blood flow in any experimental model.

Placental Blood Flow

In the chronic hypoxia, heat stress, and nutritional models, an initial reduction in uterine blood flow is an integral part of the maternal physiological response to the given stimulus. However, when flow returns to normal in the chronic hypoxia model (Kitanaka et al. 1989a), there is no evidence of fetoplacental growth restriction at term. Furthermore, in each of the flow-reduction models and control groups, there is a direct correlation between placental weight at term and the absolute rate of uteroplacental blood flow. As the placenta attains its maximum weight some 15 days prior to the onset of flow reduction, the flow reduction must initiate a flow-dependent decrease in placental weight. Likewise, the maintenance of normal substrate levels in both the maternal and fetal circulations until fetoplacental growth restriction is well advanced seems to indicate that the primary regulator is flow not substrate. However, as noted above, the variables of substrate delivery and flow rate change concomitantly in all experiments reported to date.

A Central Mechanism for Regulating the Interaction

All methods of producing a reduction in fetal growth are associated with a concomitant and usually balanced reduction in placental weight. In turn, in all the models in which it has been measured, both placental weight and fetal weight are closely correlated with a concomitant restriction in the rate of uteroplacental perfusion. Furthermore, fetal growth rate appears to be quite sensitive to minor changes in placental perfusion and/or substrate delivery, and the decrease in growth rate appears to be an initial adaptive response to a reduction in perfusion as it occurs very early on without concomitant evidence of metabolic instability. This all indicates that the placenta has a central role in mediating the interaction between substrate availability at the maternal-fetal interface and fetal growth.

A very simple interactive mechanism would be a sensor in placental tissue which responded to a decrease or increase in blood flow and/or substrate delivery by either initiating or suppressing the placental synthesis and release of a growth suppressing substance into the fetal compartment. Such a mechanism is not only physiologically sound, but its existence is supported by several lines of evidence detailed elsewhere (Clapp 1989). The most convincing data is the observation that the placental venous effluent from growth-retarded pregnancies contains a peptide which suppresses hepatocyte protein synthesis and a plasma factor which inhibits the sulphation of fetal cartilage in a dose-dependent fashion despite elevated insulin-like growth factor (IGF) II levels (Jones 1985, Jones et al. 1987). Second, hormonal profiles in the embolization model suggest that the placental production and release of steroids is responsive to changing patterns of placental bed blood flow (Clapp et al. 1982b). Third, the normal variability in fetal growth rate in late gestation appears to be superimposed on the inherent variability in the earlier growth of the placenta in control animals and in the embolization and nutritional deprivation models (Mellor 1983; Clapp 1989). Finally, the observation that fetoplacental growth restriction can be prevented by fetal substrate infusion (Charlton and Johengen 1987) indicates that the placenta must play a central role in regulating the interaction.

Summary

In summary, current data indicate that both placental and fetal weight at term are limited by the rate of maternal placental perfusion and/or substrate delivery. Oxygen per se does not appear to be a primary modulator of the interaction. Rather, it appears likely that the adequacy of the initial process of placentation itself coupled with placental-fetal signaling in response to fluctuations in flow and/or substrate availability regulates fetal growth rate.

References

Alexander G, Hales JRS, Stevens D, Donnelly JB (1987) Effects of acute and prolonged exposure to heat on regional blood flows in pregnant sheep. J Dev Physiol 9: 1–15

Bell AW, Wilkening RB, Meschia G (1987) Some aspects of placental function in chronically heat stressed ewes. J Dev Physiol 9: 17–29

Charlton V, Johengen M (1987) Fetal intravenous nutritional supplementation ameliorates the development of embolization-induced growth retardation in sheep. Pediatr Res 22: 55–61

Clapp JF (1989) Utero-placental blood flow and fetal growth. In: Sharp F, Fraser RB, Milner RDG (eds) Fetal growth. The Royal College of Obstetricians and Gynaecologists, London, p 235

Clapp JF, Szeto HH, Larrow R, Hewitt J, Mann LI (1981) Fetal metabolic response to experimental placental vascular damage. Am J Obstet Gynecol 140: 446–451

Clapp JF, McLaughlin MK, Larrow R, Farnham J, Mann LI (1982a) The uterine hemodynamic response to repetitive unilateral vascular embolization in the pregnant ewe. Am J Obstet Gynecol 82: 30%I –318

Clapp JF, Auletta FJ, Farnham J, Larrow R, Mann LI (1982b) The ovine fetoplacental endocrine response to placental damage. Am J Obstet Gynecol 144: 47–54

Clark KE, Mack CE (1987) A new model of growth retardation – fetal growth is blood flow dependent. Proceedings of teh Society of Gynecol Invest (abstr no 137)

Gilbert RD, Cummings LA, Juchau MR, Longo LD (1979) Placental diffusing capacity and fetal development in exercising or hypoxic guinea pigs. J Appl Physiol 46: 828–834
Jacobs R, Robinson JS, Owens JA, Falconer J, Webster MED (1988a) The effect of prolonged hypobaric hypoxia on growth of fetal sheep. J Dev Physiol 10: 97–112
Jacobs R, Owens JA, Falconer J, Webster MED, Robinson JS (1988b) Changes to metabolite concentration in fetal sheep subjected to prolonged hypobaric hypoxia. J Dev Physiol 10: 113–121
Jones CT (1985) Reprogramming of metabolic development by restriction of fetal growth. Biochem Soc Trans 13: 89–91
Jones CT, Lafeber HN, Price DA, Parer JT (1987) Studies on the growth of the fetal guinea pig. Effects of a reduction in uterine blood flow on plasma sulphation-promoting activity and on the concentration of insulin-like growth factors-I and -II. J Dev Physiol 9: 181–201
Kitanaka T, Gilbert RD, Longo LD (1989a) Maternal responses to long-term hypoxemia in sheep. Am J Physiol 256: R1340–1347
Kitanaka T, Alonso JG, Gilbert RD, Clemons GK, Longo LD (1989b) Fetal responses to long-term hypoxemia in sheep. Am J Physiol 256: R1348–1354
Lichty JA, Ting RY, Burns PD, Dyar E (1957) Studies of babies born at high altitude. I. Relation of altitude to birth weight. Am Med Assoc J Dis Child 93: 666–678
Mellor D (1983) Nutritional and placental determinants of foetal growth rate in sheep and consequences for the newborn lamb. Br Vet J 139: 307–324
Morriss FH, Rosenfeld CR, Crandell SS, Adcock EW (1980) Effects of fasting on uterine blood flow and substrate uptake in sheep. J Nutr 110: 2433–2443
Owens JA, Falconer J, Robinson JS (1986) Effect of restriction of placental growth on umbilical and uterine blood flows. Am J Physiol 250: R427–434
Robinson JS, Falconer J, Owens JA (1985) Intrauterine growth retardation: clinical and experimental. Acta Paediatr Scand Suppl 319: 135–142

*Regulation of Perinatal Pulmonary Blood Flow: Role of Oxygen**

M.A. HEYMANN

Introduction

In the fetus, respiratory gas exchange occurs in the placenta and not in the lung. Consequently, relative to the normally high pulmonary vascular resistance, pulmonary flow is low (about 35 ml/min per kilogram fetal body weight in near-term fetal lambs). With the onset of pulmonary ventilation at birth, pulmonary vascular resistance falls rapidly, and pulmonary blood flow increases approximately eightfold to about 300 ml/min per kilogram body weight shortly after birth.

Many factors are thought to be involved in the regulation of pulmonary blood flow in the critical period. The O_2 environment is an important factor, as is the local release of vasoactive substances, which may play a more direct role in both the immediate postnatal rapid pulmonary vasodilatation and the subsequent maintenance of pulmonary vasodialatation. Perhaps even more important is the interaction between O_2 and various vasoactive substances such as bradykinin and the eicosanoids (prostaglandins, PGs, thromboxane, and leukotrienes, LTs), which are metabolic products of arachidonic acid (AA) metabolism.

Role of Oxygen

The level of O_2 to which the small pulmonary arteries (resistance vessels) are exposed plays a role in controlling pulmonary blood flow both in utero and after birth (Heymann and Hoffman 1986; Lewis et al. 1976; Rudolph et al. 1977; Rudolph and Yuan 1966). Under normal baseline conditions in the fetal lamb, pulmonary arterial blood O_2 is about 17 torr. Reducing PaO_2 to similar levels in normal newborn calves markedly increases pulmonary vascular resistance (Rudolph and Yuan 1966), and conversely, increasing fetal O_2 above the normally low levels increases pulmonary blood flow (Heymann et

* This work was supported in part by U.S. Public Health Service Program Project Grant HL 24056 and by U.S. Public Health Service Grant HL 40473 from the National Institutes of Health

al. 1969). Probably, therefore, the low O_2 environment in the fetus helps to maintain pulmonary vasoconstriction, increased pulmonary vascular resistance, and, thereby, a low pulmonary blood flow; furthermore, the postnatal increase in O_2 is probably involved in the fall in pulmonary vascular resistance and the increase in pulmonary blood flow.

The exact mechanisms responsible for the low O_2-related pulmonary vasoconstriction in the fetus (and also after birth) are not clearly established. When the O_2 environment of a single fetal lung in situ is changed, either by ventilating or perfusing that lung independently without changing the O_2 environment of the other lung, vascular resistance changes are produced only in the perfused or ventilated lung. This suggests that the O_2-related changes in the isolated lung are somehow locally mediated (Campbell et al. 1967).

Although gaseous expansion or ventilation of fetal lungs without an increase in PaO_2 reduces pulmonary vascular resistance, a further substantial fall occurs when O_2 is added to the ventilating gas mixture (Cassin et al. 1964). This vasodilating effect of O_2 has been further substantiated by exposing fetal lambs to hyperbaric oxygenation without ventilating their lungs (Heymann et al. 1969). It is not known whether the increased O_2 directly relaxes pulmonary vascular smooth muscle or whether the rise in O_2 environment results in the secondary release of a substance or substances that then actively dilate the pulmonary circulation (either directly or perhaps even through the further release of a different vasoactive substance) or inhibit active vasoconstriction; alternately, the increase in O_2 might directly inhibit the production of substances that actively constrict the small pulmonary arteries.

Role of Vasoactive Substances

Many vasoactive substances, administered in exogenous form, affect the pulmonary circulation, but whether they play a physiologic role in maintaining the high fetal pulmonaty vascular resistance or in the postnatal vasodilatation is not known. For example, angiotensin II (AII) has been suggested to play some role in the pulmonary vasoconstrictor response to hypoxemia in adults (Berkov 1974); however, in the fetus, blockade of AII activity with saralasin did not affect either baseline pulmonary vascular resistance or the response to further induced hypoxemia (Hyman et al. 1975). Conversely, the increase in AII concentration after birth may play a role in postnatal pulmonary vasodilatation by stimulating prostaglandin I_2 (PGI_2) release (Omini et al. 1983) (see below).

In adults, PGs do not circulate in plasma and, where involved in vascular regulation, they are locally synthesized; however, in the fetus, measureable amounts of PGs are present in plasma (Challis and Patrick 1980; Mitchell et al. 1978). Although these PGs are probably derived mainly from the placenta, fetal vasculature, including the pulmonary arteries, can produce PGs (Challis and Patrick 1980; Powell and Solomon 1977; Terragno et al.

1980). In both the fetus and newborn, exogenous PGs or LTs have diverse effects, and inhibitors of PG or LT synthesis or blockers of LT activity likewise affect the circulation. This suggests an intrinsic role for endogenous PGs and LTs in circulatory, and particularly pulmonary circulatory, homeostasis during this period of development.

Several metabolites of AA are involved in the physiologic regulation of the perinatal pulmonary circulation. The PGs initially received most attention; however, more recent studies indicate that the LTs also may play a role, perhaps interacting with PGs in the regulation of pulmonary blood flow in the fetus in particular and also possibly during hypoxic stress postnatally.

Exogenously administered PGs have potent vasoactivity and, although they also have systemic vascular effects, PGE_2 and particularly PGI_2 are pulmonary vasodilators (Cassin et al. 1981; Leffler and Hessler 1979). The fact that exogenous PGs exert systemic effects does not exclude the possibility that under physiologic conditions, production and action accurs locally in the lungs without systemic effects. There is now convincing evidence that PGI_2 is involved in physiologic postnatal pulmonary vasodilatation. Mechanical stimulation, such as stretching of the lungs, leads to PG or PGI_2 production (Edmonds et al. 1969; Gryglewski et al. 1978), and ventilation of fetal lungs, even without changing the O_2 environment, increases pulmonary venous plasma concentrations of the stable metabolite of PGI_2, 6-keto-prostaglandin $F_{1\alpha}$ (Leffler et al. 1980, 1984a, b). The 6-keto-PG $F_{1\alpha}$ concentrations fall over several hours, further suggesting a physiologic role of PGI_2 in the immediate postnatal period.

Additional support for the role of PGI_2 is provided by studies showing that AII, which increases in concentration immediately after birth, induces the release of PGI_2 by the lung (Dusting 1981). In vitro studies with pulmonary vascular endothelial cells in culture have shown direct stimulation of PGI_2 release by several vasoactive substances, including AII (McIntyre et al. 1985), and the newborn lung, when stimulated with AII to produce PGI_2, will do so at lower concentrations of AII than the adult or fetal lung (Omini et al. 1983). It has also been established that shortly after birth, the conversion of angiotensin I (AI) to AII by angiotensin converting enzyme (ACE) increases dramatically; this is related to the increase in surface area of the pulmonary vascular bed that occurs with the increase in pulmonary blood flow (Lister et al. 1984; Pitt and Lister 1984).

To fit the hypothesis presented here, the latter phenomenon must be brought about by a mechanism not related directly to this particular cascade. This is of course possible because, as outlined, many different mechanisms can produce pulmonary vasodilatation, and at least several are likely involved consecutively or together in effecting the changes at and after birth. AII thus produced within a pulmonary vascular endothelial cell could then stimulate AA metabolism to produce PGI_2, which in turn would relax pulmonary vascular smooth muscle. The action of PGI_2 on smooth muscle is most likely generated through a specific membrane receptor stimulating a coupling protein (G protein s), then adenylate cyclase activation, cyclic adenosine monophosphate (cAMP) production, and eventually muscle relaxation through protein phosphorylation.

Perhaps even more important than AII in stimulating pulmonary PGI_2 production is bradykinin, which is produced by the lung with the onset of oxygenation and which, when administered exogenously, produces potent pulmonary vasodilatation (see below). Although its action could be direct or mediated through the release of endothelial-derived relaxing factor (EDRF), bradykinin also is known to be a potent stimulator of PGI_2 production (McIntyre et al. 1985).

The role of vasodilating PGs in the immediate postnatal period has been further substantiated when cyclo-oxygenase activity, and thereby PG production, was inhibited just before the onset of ventilation (Cassin 1980; Leffler et al. 1978). The fall in pulmonary vascular resistance with ventilation has at least two separate phases: an initial rapid fall within 30–60 s, followed by a slower fall over about 15 min. This latter, second phase was inhibited or attenuated by pretreatment with indomethacin. Although not defining the specific PGs involved, nor pointing to any mechanism that might be responsible for inducing their production, these studies indicate clearly that PGs play a role.

The 5′-lipoxygenase products of AA metabolism, LTs, are now thought to be responsible, at least in part, for mediating hypoxic pulmonary vasoconstriction in adults (Ahmed and Oliver 1983; Morganroth et al. 1984). We showed attenuation of hypoxia-induced pulmonary vasoconstriction in newborn lambs with prior LT antagonism (Schreiber et al. 1985), and, because of the normally low fetal PaO_2, we considered that baseline LT production could also be involved in regulating fetal pulmonary vasoconstrictor tone. In fetal lambs, end organ antagonism of LT effect with FPL 57231 (Fisons, Loughborough, UK) or LT synthesis inhibition with U 60257B (Upjohn, Kalamazoo, Michigan, USA) indeed increased pulmonary blood flow to levels approaching those expected with normal ventilation after birth (Le Bidois et al. 1987; Soifer et al. 1985). These studies strongly suggest a physiologic role for LTs in maintaining pulmonary vasoconstriction and thereby a low pulmonary blood flow in the fetus. Further studies have shown that these effects are not produced through the secondary stimulation of thromboxane production (Clozel et al. 1985) and that LTD_4 is the LT actively involved (Schreiber et al. 1987).

Eicosanoid regulation of the perinatal pulmonary circulation, therefore, reflects a balance between factors producing active pulmonary vasoconstriction in the fetus and those leading to pulmonary vasodilatation with ventilation. The dramatic increase in pulmonary blood flow after birth could reflect a major shift from active vasoconstriction to active vasodilatation. AA or even phospholipid metabolism may change from the 5′-lipoxygenase pathway (LTs) in the low fetal O_2 environment to the cyclo-oxygenase pathway (PGs), either in response to mechanical stimulation (lung expansion) or more likely to the higher O_2 environment after birth.

Bradykinin, a potent vasoactive peptide in adults, also produces marked pulmonary vasodilatation when infused into fetal lambs (Campbell et al. 1968). Because in fetal lambs both the concentration of bradykinin precursor, kininogen, falls and bradykinin is released from the lungs following fetal oxygenation, either by ventilation with air or during hyperbaric oxygenation (Heymann et al. 1969), we considered that bradykinin could also play a physiologic role in the immediate postnatal pulmonary circulatory changes following the initiation of ventilation. It is unclear whether the effects of bradykinin are direct or indirect, but still receptor-mediated, or what proportion of the effect on the pulmonary circulation is via production of PGs or of another vasoactive substance, such as EDRF. Addition of bradykinin to endothelial cells in culture increases the media concentration of 6-keto-prostaglandin $F_{1\alpha}$ (Hong 1980; McIntyre et al. 1985). Thus, bradykinin-induced PGI_2 production clearly is one potential mechanism for perinatal pulmonary vasodilatation, and our recent studies of bradykinin-induced pulmonary vasodilatation before and with cyclo-oxygenase inhibition in fetal lambs show that only 30%–40% of the effect of bradykinin is PGI_2 mediated (Frantz et al. 1989).

Many pharmacologic or physiologic vasoactive substances previously thought to have direct effects on pulmonary vascular smooth muscle, have new been shown to act through the specific release from pulmonary vascular endothelium of a secondary substance with the vasoactive properties (Gryglewski et al. 1988; Peach et al. 1985). A prime example of this is the well-described effect of acetylcholine, which does not produce pulmonary vasodilatation directly, but rather by acting on a muscarinic receptor to stimulate the production of EDRF (Furchgott 1983). Although still unproven, similar mechanisms could be involved in perinatal pulmonary vasodilatation. For example, vascular endothelial cell EDRF production can be stimulated by a diverse group of substances including bradykinin, acetylcholine, AII, arginine vasopressin, catecholamines, histamine, adenosine diphosphate, etc. Oxygen-induced bradykinin production could, therefore, act through such a pathway and be responsible for part or all of the remaining 60%–70% of the bradykinin-related perinatal pulmonary vasodilatation.

Bradykinin, therefore, could produce its postnatal pulmonary vasoilator effect by three different mechanisms: first, a direct smooth muscle effect; second, through the release of PGI_2, as indicated above; and third, through the stimulation of EDRF production. Our study (Frantz et al. 1989) defined an interaction between bradykinin and PGI_2; however, part of the effect of bradykinin was clearly independent of PG production because indomethacin only partially blocked the pulmonary vasodilatation even though it completely inhibited PGI_2 production (measured as the stable metabolite 6-ket-prostaglandin $F_{1\alpha}$). The mechanism for the perinatal pulmonary vasodilatation by bradykinin independent of PGI_2 production has not yet been fully established. In adult arteries, bradykinin-related relaxation required an intact endothelium and is produced by EDRF (Furchgott 1983). This very likely is the mechanism in our lambs as well, because in preliminary studies we have been able to attenuate the effect of bradykinin with methylene blue, a cyclic guanosine monophosphate (cGMP) inhibitor.

We propose that bradykinin, by a mechanism independent of PGI_2 production, may be responsible for the initial decrease in pulmonary vascular resistance, and that PGI_2 may be responsible for the subsequent, slower decrease. Such a role for bradykinin in this sequence is consistent with several other observations. We have shown that bradykinin precursor concentrations (kininogen) increased with gestational age in fetal lambs, and that these concentrations fell and bradykinin appeared in left atrial blood shortly after oxygenation (Heymann et al. 1969). The cascade is initiated by an increase in O_2 leading to endothelial production of bradykinin. This could act directly on smooth muscle, or more likely, could initiate one of two possible additional mechanisms. The first would involve bradykinin-mediated stimulation of endothelial cell AA metabolism to produce PGI_2, which then would produce cAMP-mediated smooth muscle relaxation. The second would involve a bradykinin-stimulated endothelial cell release of EDRF, which also could act on smooth muscle to produce relaxation. This latter bradykinin effect is receptor-initiated; however, the exact mechanism by which EDRF produces relaxation is not fully established, although guanylate cyclase-related cGMP stimulation is known to be involved.

Although many of the phenomena described have been shown to occur as independent and unrelated events, many of the linkages proposed are pure speculation and, therefore, are theoretic and unproven. The pieces of the puzzle are starting to fit together, but there are many gaps, and some of the pieces may in fact be in the wrong places. It remains for further study to make the necessary changes.

References

Ahmed T, Oliver W Jr (1983) Does slow-reacting substance of anaphylaxis mediate hypoxic pulmonary vasoconstriction? Am Rev Respir Des 127: 566–571

Berkov S (1974) Hypoxic pulmonary vasoconstriction in the rat: the necessary role of angiotensin II. Circ Res 35: 256–261

Campbell AGM, Cockburn F, Dawes GS et al. (1967) Pulmonary vasoconstriction in asphyxia during cross-circulation between twin foetal lambs. J Physiol (Lond) 192: 111–122

Campbell AGM, Dawes GS, Fishman AP et al. (1968) The release of a bradykinin-like pulmonary vasodilator substance in foetal and newborn lambs. J Physiol (Lond) 195: 83–96

Cassin S (1980) Role of prostaglandins and thromboxanes in the control of the pulmonary circulation in the fetus and newborn. Semin Perinatol 4: 101–107

Cassin S, Dawes GS, Ross BB (1964) Pulmonary blood flow and vascular resistance in immature foetal lambs. J Physiol (Lond) 171: 80–89

Cassin S, Winikor I, Tod M et al. (1981) Effects of prostacyclin on the fetal pulmonary circulation. Pediatr Pharmacol 1: 197–207

Challis JRG, Patrick JE (1980) The production of prostaglandins and thromboxanes in the feto-placental unit and their effects on the developing fetus. Semin Perinatol 4: 23–33

Clozel M, Clyman RI, Soifer SJ et al. (1985) Thromboxane is not responsible for the high pulmonary vascular resistance in fetal lambs. Pediatr Res 19: 1254–1257

Dusting GJ (1981) Angiotensin-induced release of a prostacyclin-like substance from the lungs. J Cardiovasc Pharmacol 3: 197–206

Edmonds JF, Berry E, Wyllie JH (1969) Release of prostaglandins by distension of the lungs. Br J Surg 56: 622–623

Frantz E, Soifer SJ, Clyman RI et al. (1989) Bradykinin produces pulmonary vasodilation in fetal lambs: role of prostaglandin production. J Appl Physiol 67: 1512–1517

Furchgott RF (1983) Role of endothelium in responses of vascular smooth muscle. Circ Res 53: 557–573

Gryglewski RJ, Korbut R, Ocetkiewicz A (1978) Generation of prostacyclin by lungs in vivo and its release into the arterial circulation. Nature 273: 765–767

Gryglewski RJ, Botting RM, Vane JR (1988) Mediators produced by the endothelial cell. Hypertension 12: 530–548

Heymann MA, Rudolph AM, Nies AS et al. (1969) Bradykinin production associated with oxygenation of the fetal lamb. Circ Res 25: 521–534

Heymann MA, Hoffman JIE (1986) Pulmonary circulation in the perinatal period. In: Thibeault DW, Gregory GA (eds) Neonatal pulmonary care. Appleton-Century-Crofts, Norwalk, pp 149–174

Hong SL (1980) Effect of bradykinin and thrombin on prostacyclin synthesis in endothelial cells from calf and pig aorta and human umbilical cord vein. Thromb Res 18: 787–795

Hyman A, Heymann MA, Levin DL et al. (1975) Angiotensin is not the mediator of hypoxia-induced pulmonary vasoconstriction in fetal lambs. Circulation 52: II–132 (abstr)

Le Bidois J, Soifer SJ, Clyman RI et al. (1987) Piriprost" a putative leukotriene synthesis inhibitor increases pulmonary blood flow in fetal lambs. Pediatr Res 22: 350–354

Leffler CW, Hessler JR (1979) Pulmonary and systemic vascular effects of exogenous prostaglandin I in fetal lambs. Eur J Pharmacol 54: 37–42

Leffler CW, Tyler TL, Cassin S (1978) Effect of indomethacin on pulmonary vascular response to ventilation of fetal goats. Am J Physiol 234: H346–H351

Leffler CW, Hessler JR, Terragno NA (1980) Ventilation induced release of prostaglandin-like material from fetal lungs. Am J Physiol 238: H282–H286

Leffler CW, Hessler JR, Green RS (1984a) Mechanism of stimulation of pulmonary prostacyclin synthesis at birth. Prostaglandins 28: 877–887

Leffler CW, Hessler JR, Green RS (1984b) The onset of breathing at birth stimulates pulmonary vascular prostacyclin synthesis. Pediatr Res 18: 938–942

Lewis AB, Heymann MA, Rudolph AM (1976) Gestational changes in pulmonary vascular responses in fetal lambs in utero. Circ Res 39: 536–541

Lister G, Kopf GS, Pitt BR (1984) Functional assessment of the pulmonary microcirculation during postnatal development. Pediatr Pharmacol 4: 85–99

McIntyre TM, Zimmerman GA, Satoh K et al. (1985) Cultured endothelial cells synthesize both platelet-activating factor and prostacyclin in response to histamine, bradykinin, and adenosine triphosphate. J Clin Invest 76: 271–280

Mitchell MD, Flint AP, Bibby J et al. (1978) Plasma concentrations of prostaglandins during late human pregnancy: Influence of normal and preterm labor. J Clin Endocrinol Metab 46: 947–951

Morganroth ML, Reeves JT, Murphy RC et al. (1984) Leukotriene synthesis and receptor blockers block hypoxic pulmonary vasoconstriction. J Appl Physiol 56: 1340–1346

Omini C, Vigano T, Marini A, et al. (1983) Angiotensin II: a releaser of PGI_2 from fetal and newborn rabbit lungs. Prostaglandins 25: 901–910

Peach MJ, Singer HA, Loeb AL (1985) Mechanisms of endothelium-dependent vascular smooth muscle relaxation. Biochem Pharmacol 34: 1867–1874

Pitt BR, Lister G (1984) Kinetics of pulmonary angiotensin-converting enzyme activity in conscious developing lambs. J Appl Physiol 57: 1158–1166

Powell WS, Solomon S (1977) Formation of 6-oxoprostaglandin $F_{1\alpha}$ by arteries of the fetal calf. Biochem Biophys Res Commun 75: 815–822
Rudolph AM, Yuan S (1966) Response of the pulmonary vasculature to hypoxia and H^+ ion concentration changes. J Clin Invest 45: 399–411
Rudolph AM, Heymann MA, Lewis AB (1977) Physiology and pharmacology of the pulmonary circulation in the fetus and newborn. In: Hodson WA (ed) Lung biology in health and disease. Development of the lung. Dekker, New York, pp 497–523
Schreiber MD, Heymann MA, Soifer SJ (1985) Leukotriene inhibition prevents and reverses hypoxic pulmonary vasoconstriction in newborn lambs. Pediatr Res 19: 437–441
Schreiber MD, Heymann MA, Soifer SJ (1987) The differential effects of leukotriene C_4 and D_4 on the pulmonary and systemic circulations in newborn lambs. Pediatr Res 21: 176–182
Soifer SJ, Schreiber MD, Loitz RD et al. (1985) The effect of leukotriene inhibition on the perinatal pulmonary circulation in the lamb. In: Jones CT (ed) The physiologic development of the fetus and newborn. Academic, London, pp 451–455
Terragno NA, Terragno A, McGiff JC (1980) Role of prostaglandins in blood vessels. Semin Perinatol 4: 85–90

Abstracts

Cardiovascular Responses to Acute Asphyxia in Carotid Sinus Denervated and Intact Fetal Sheep Near Term

A. JENSEN and M.A. HANSON

Acute fetal asphyxia from occlusion of uterine blood flow produces a rapid fetal peripheral vasoconstriction, a decrease in heart rate (HR) and an increase in arterial blood pressure (ABP) (Jensen et al. 1987). The rapidity of these responses suggests that they involve reflexes elicited from the peripheral chemoreceptors.

To test this we studied three fetuses after bilateral division of the carotid sinus nerves (CSNX) and five intact fetuses (SHAM) at 130±2 days gestation. Instrumentation for placement of catheters was as in previous

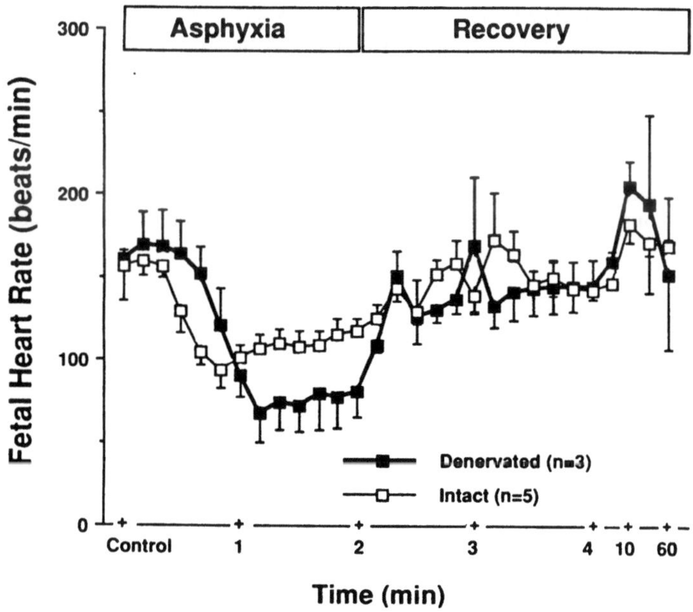

Fig. 1. Fetal heart rate. Change occurring during and after 2 min of asphyxia in intact and carotid sinus denervated fetal sheep near term. Our results provide evidence for the involvement of chemo- and baroreflexes in the fetal circulatory and sympathoadrenal resonses to acute asphyxia. A lack of chemoreflex compensation in CSNX prevents peripheral vasoconstriction, and SO_2 falls to lower levels than in SHAM. The greater variability of heart rate and arterial blood pressure in CSNX is probably due to lack of baroreflexes.

experiments (Jensen et al., 1987). On the 3rd postoperative day, asphyxia was produced by arrest of uterine blood flow by occluding the maternal descending aorta below the origin of the renal arteries for 2 min. At 0, 1, 2, 3, 4, and 30 min preductal arterial blood was withdrawn for determination of blood gases and O_2 saturation (SO_2), acid base status, adrenaline (Adr) and noradrenaline (NA) levels. During the control period there were no differences between CSNX and SHAM, except that ABP was significantly higher in CSNX than in SHAM (49.6±0.7 vs. 42.6±2.2). During asphyxia, Heart rate fell and arterial blood pressure rose in both groups, but the changes were delayed by 20 s and 60 s, in CSNX and SHAM, respectively. After the delay, however, heart rate fell more rapidly and to lower levels in CSNX (80±15) than in SHAM (117±8) (Fig. 1). The rise in arterial bood pressure was also steeper in CSNX after the delay, peak arterial blood pressure being reached in both groups approximately 1 min after termination of the asphyxia. Both heart rate and arterial blood pressure were more variable in CSNX than SHAM. Adr rose by about 3 orders of magnitude in CSNX after 2 min of asphyxia, but significantly less in SHAM. There were no differences in the time course or size of the NA increase. Lastly, SO_2 fell significantly more in CSNX than in SHAM.

Acknowledgements. This work was supported by the Deutsche Forschungsgemeinschaft Je 108/4–2 and the Wellcome Trust.

Reference

Jensen A, Hohmann M, Künzel W (1987) Dynamic changes in organ blood flow and oxygen consumption during acute asphyxia in fetal sheep. J Dev Physiol 9: 543–559

Subject Index

MIX
Papier aus verantwortungsvollen Quellen
Paper from responsible sources
FSC® C105338

If you have any concerns about our products,
you can contact us on
ProductSafety@springernature.com

In case Publisher is established outside the EU,
the EU authorized representative is:
Springer Nature Customer Service Center GmbH
Europaplatz 3, 69115 Heidelberg, Germany

Printed by Libri Plureos GmbH
in Hamburg, Germany